ELEMENTARY PARTICLES AND THE LAWS
OF PHYSICS

Sketch of P. A. M. Dirac by R. P. Feynman

ELEMENTARY PARTICLES AND THE LAWS OF PHYSICS

THE 1986 DIRAC MEMORIAL LECTURES

RICHARD P. FEYNMAN
California Institute of Technology

and

STEVEN WEINBERG
University of Texas, Austin

Lecture notes compiled by
Richard MacKenzie and Paul Doust

CAMBRIDGE
UNIVERSITY PRESS

Published by the Press Syndicate of the University of Cambridge
The Pitt Building, Trumpington Street, Cambridge CB2 1RP
40 West 20th Street, New York, NY 10011-4211, USA
10 Stamford Road, Oakleigh, Melbourne 3166, Australia

First published 1987
Reprinted 1988, 1989, 1991, 1993, 1994, 1995

Printed in the United States of America

British Library Cataloging in Publication Data
Feynman, Richard P.
Elementary particles and the laws of
physics: the 1986 Dirac memorial lectures
1. Quantum field theory
I. Title   II. Weinberg, Steven
530.1'2   QC174.45

Library of Congress Cataloging in Publication Data
Feynman, Richard Phillips
Elementary particles and the laws of physics: the 1986 Dirac
memorial lectures/Richard P. Feynman and Steven Weinberg;
lecture notes compiled by Richard MacKenzie and Paul Doust
Includes bibliographies
1. Particles (Nuclear physics) 2. Quantum theory 3. Relativity
(Physics) 4. Dirac, P.A.M. (Paul Adrien Maurice), 1902-
I. Weinberg, Steven. II. Title. III. Title: Dirac memorial lectures
QC793.28.F49   1987
539.7'21-dc19

ISBN 0-521-34000-4 hardback

CONTENTS

FOREWORD

John C. Taylor

University of Cambridge

Paul Dirac was one of the finest physicists of this century. The development of quantum mechanics began at the turn of the century, but it was Dirac who, in 1925 and 1926, brought the subject to its definitive form, creating a theory as compelling as Newton's mechanics had been.

Dirac immediately set about reconciling the quantum theory with Einstein's special theory of relativity (of 1905). The nature of the marriage between these two marvellous theories, and the fruits of that union, have been the constant preoccupation of fundamental physics from 1925 to the present day. Dirac contributed more than anyone else to this crucial enterprise, including in 1930 the prediction of the existence of antimatter.

Dirac died in 1984, and St John's College, Cambridge (Dirac's college), very generously endowed an annual lecture to be held at Cambridge University in Dirac's memory. The first two Dirac Lectures, printed here, are contrasting variations on Dirac's theme of the union of quantum theory and relativity.

Richard Feynman, in the years since the Second World War, did more than anyone else to evolve Dirac's relativistic quantum theory into a general and powerful method of making physical predictions about the interactions of particles and radiation. His work complements Dirac's in a remarkable way. His style of doing physics has been vastly influential. His lecture here, which gives some flavour of that style, expounds the physical reality underlying Dirac's prediction of antimatter.

The crowning achievement to date of the relativistic quantum theory has been the unification of electricity and magnetism on the one hand (themselves unified by Maxwell a century ago) with the weak forces of radioactive decay on the other. Steven Weinberg is one of the chief authors of this unification, in work which predicted the existence and properties of new particles (weighing as much as heavy atoms), which were subsequently triumphantly produced, precisely as predicted, at the

European laboratory CERN in Geneva in 1983. This echoed Dirac's prediction, half a century earlier, of the positron and its subsequent discovery, though the energy necessary to produce a positron was 100 000 times less.

In his lecture, Weinberg shows how tightly quantum theory and relativity together constrain the laws of Nature, and he speculates how Einstein's theory of gravitation (of 1915) will be reconciled with quantum theory.

We in Cambridge were fortunate that these two leading physicists agreed to commemorate Dirac by coming to lecture here. They drew audiences of several hundred undergraduates and graduates, some of them physicists, some not. Both Feynman and Weinberg have been concerned to explain physics to nonspecialists*, and we hope that this volume too will interest a wide readership.

Dirac stated his philosophy of physics in the sentence 'physical laws should have mathematical

---

* R. P. Feynman, R. B. Leighton and M. Sands (1963). *Lectures in Physics*, vols. 1–3. Addison-Wesley.

R. P. Feynman (1985). *QED*. Princeton University Press.

S. Weinberg (1978). *The First Three Minutes*. Fontana. (Previously published in 1977 by Deutsch.)

S. Weinberg (1983). *The Discovery of Subatomic Particles*. Scientific American Library.

beauty'[†]. Dirac, Feynman and Weinberg have each made beautiful theories which have been spectacularly upheld in experimental tests. But the experiments, outside the scope of these Lectures, are another story.

John Taylor
September 1987

[†] See R. H. Dalitz and R. Peierls (1986). In *Biographical Memoirs of Fellows of the Royal Society*, vol. 32, pp 137–86. Royal Society.

THE REASON
FOR ANTIPARTICLES

Richard P. Feynman

The title of this lecture is somewhat incomplete because I really want to talk about two subjects: first, why there are antiparticles, and, second, the connection between spin and statistics. When I was a young man, Dirac was my hero. He made a breakthrough, a new method of doing physics. He had the courage to simply guess at the form of an equation, the equation we now call the Dirac equation, and to try to interpret it afterwards. Maxwell in his day got his equations, but only in an enormous mass of 'gear wheels' and so forth.

I feel very honored to be here. I had to accept the invitation, after all he was my hero all the time, and it is kind of wonderful to find myself giving a lecture in his honor.

Dirac with his relativistic equation for the electron was the first to, as he put it, wed quantum mechanics and relativity together. At first he

1

Paul Dirac   Richard Feynman

thought that the spin, or the intrinsic angular momentum that the equation demanded, was the key, and that spin was the fundamental consequence of relativistic quantum mechanics. However, the puzzle of negative energies that the equation presented, when it was solved, eventually showed that the crucial idea necessary to wed quantum mechanics and relativity together was the existence of antiparticles. Once you have that idea, you can do it for any spin, as Pauli and Weisskopf proved, and therefore I want to start the other way about, and try to explain why there must be anti-

2

particles if you try to put quantum mechanics with relativity.

Working along these lines will permit us to explain another of the grand mysteries of the world, namely the Pauli exclusion principle. The Pauli exclusion principle says that if you take the wave-function for a pair of spin $\frac{1}{2}$ particles and then interchange the two particles, then to get the new wavefunction from the old you must put in a minus sign. It is easy to demonstrate that if Nature was nonrelativistic, if things started out that way then it would be that way for all time, and so the problem would be pushed back to Creation itself, and God only knows how that was done. With the existence of antiparticles, though, pair production of a particle with its antiparticle becomes possible, for example with electrons and positrons. The mystery now is, if we pair produce an electron and a positron, why does the new electron that has just been made have to be antisymmetric with respect to the electrons which were already around? That is, why can't it get into the same state as one of the others that were already there? Hence, the existence of particles and antiparticles permits us to ask a very simple question: if I make two pairs of electrons and positrons and I compare the amplitudes for when they annihilate directly or for when they exchange before they annihilate, why is there a minus sign?

All these things have been solved long ago, in a beautiful way which is simplest in the spirit of Dirac with lots of symbols and operators. I am going to go further back to Maxwell's 'gear wheels' and try to tell you as best I can a way of looking at these things so that they appear not so mysterious. I am adding nothing to what is already known; what follows is simply exposition. So here we go as to how things work–first, why there must be anti-particles.

RELATIVITY AND ANTIPARTICLES

In ordinary nonrelativistic quantum mechanics, if you have a disturbing potential U acting on a particle which is initially in a state ϕ_0, then the state will be different after the disturbance. Up to a phase factor and taking $\hbar = 1$, the amplitude to end up in a state χ is given by the projection of χ onto $U\phi_0$. In fact, we have:

$$\text{Amp}_{\phi_0 \to \chi} = -i \int d^3\mathbf{x}\, \chi^* U \phi_0 = -i \langle \chi | U | \phi_0 \rangle. \quad (1)$$

The expression $\langle \chi | U | \phi_0 \rangle$ is Dirac's elegant bra and ket notation for amplitudes, although I will not use it much here. I will suppose though that

The reason for antiparticles

this formula is true when we go to relativistic quantum mechanics.

Now suppose that there are two disturbances, one at a time t_1 and another at a later time t_2, and we would like to know what the amplitude is for the second disturbance to restore the particle to its original state ϕ_0. Call the first disturbance U_1 at time t_1, and the second U_2 at time t_2. We will need to express the successive operations of: the disturbance U_1, evolution from time t_1 to t_2, and the disturbance U_2–this we will do using perturbation theory. Of course, the simplest thing that could happen is that we go straight from ϕ_0 to ϕ_0 direct, with amplitude $\langle \phi_0 | \phi_0 \rangle = 1$. This is the leading order term of the perturbation expansion. It is the next to leading order term that corresponds to the disturbance U_1 putting the state ϕ_0 into some intermediate state ψ_m of energy E_m, which lasts for time $(t_2 - t_1)$, before the other disturbance U_2 converts back to ϕ_0. All possible intermediate states must be summed over. The total amplitude for the state ϕ_0 to end up in the same state ϕ_0 is then:

$$\text{Amp}_{\phi_0 \to \phi_0} = 1 - \sum_m \langle \phi_0 | U_2(\mathbf{x}_2) | \psi_m \rangle$$

$$\times \exp(-i E_m(t_2 - t_1)) \langle \psi_m | U_1(\mathbf{x}_1) | \phi_0 \rangle. \quad (2)$$

RICHARD P. FEYNMAN

(I have assumed, for simplicity, that there is no first order amplitude to go from ϕ_0 to ϕ_0; that is, that $\langle \phi_0 | U_1 | \phi_0 \rangle = 0$ and $\langle \phi_0 | U_2 | \phi_0 \rangle = 0$.) If we use plane waves for the intermediate states ψ_m and expand out the amplitudes $\langle \phi_0 | U_2 | \psi_m \rangle$ and $\langle \psi_m | U_1 | \phi_0 \rangle$, we see that

$$\text{Amp}_{\phi_0 \to \phi_0} = 1 - \int d^3\mathbf{x}_1 \, d^3\mathbf{x}_2 \int \frac{d^3\mathbf{p}}{(2\pi)^3 2E_p} b^*(\mathbf{x}_2)$$

$$\times \exp\left\{ -i\left[E_p(t_2 - t_1) \right. \right.$$

$$\left. \left. - \mathbf{p} \cdot (\mathbf{x}_2 - \mathbf{x}_1) \right] \right\} a(\mathbf{x}_1). \quad (3)$$

Here

$$a(\mathbf{x}_1) = U_1(\mathbf{x}_1)\phi_0(\mathbf{x}_1)\sqrt{(2E_p)},$$

$$b(\mathbf{x}_2) = U_2(\mathbf{x}_2)\phi_0(\mathbf{x}_2)\sqrt{(2E_p)},$$

and $E_p = \sqrt{(p^2 + m^2)}$ for a particle of mass m. These E_p factors are arranged just to make the relativistic properties more apparent, as $d^3\mathbf{p}/(2\pi)^3 2E_p$ is an invariant momentum density. The process can be written pictorially as in Fig. 1.

We are going to study some special cases of the above formula. The way I am going to do it is first

6

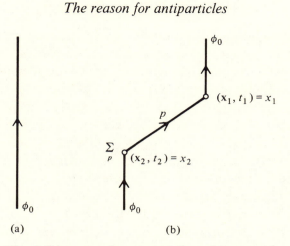

Fig. 1 Diagrammatic representation of two contributions to the amplitude for the transition $\phi_0 \to \phi_0$. (a) Direct; (b) indirect.

to examine some very simple examples and then proceed a little more generally. Hopefully you will understand the simple examples, because if you do you will understand the generalities at once–that's the way I understand things anyway.

In the indirect amplitude the particle is scattered from $\mathbf{x}_1$ to $\mathbf{x}_2$ and the intermediate states are particles with momentum $\mathbf{p}$ and energy E_p. We are going to suppose something: that all the energies are positive. If the energies were negative we know that we could solve all our energy problems by

7

dumping particles into this pit of negative energy and running the world with the extra energy.

Now here is a surprise: if we evaluate the amplitude for any $a(\mathbf{x}_1)$ and $b(\mathbf{x}_2)$ (we could even arrange for $a(\mathbf{x}_1)$ and $b(\mathbf{x}_2)$ to depend on $\mathbf{p}$) we find that it cannot be zero when $\mathbf{x}_2$ is outside the light cone of $\mathbf{x}_1$. This is very surprising: if you start a series of waves from a particular point they cannot be confined to be inside the light cone if all the energies are positive. This is the result of the following mathematical theorem:

If a function $f(t)$ can be Fourier decomposed into positive frequencies only, i.e. if it can be written

$$f(t) = \int_0^\infty e^{-i\omega t} F(\omega)\, d\omega, \qquad (4)$$

then f cannot be zero for any finite range of t, unless trivially it is zero everywhere. The validity of this theorem depends on $F(\omega)$ satisfying certain properties, the details of which I would prefer to avoid.

You may be a bit surprised at this theorem because you know you can take a function which is zero over a finite range and Fourier analyze it, but then you get positive *and* negative frequencies. I am insisting that the frequencies be positive only.

To apply this theorem to the case at hand, we fix x_1 and x_2 and rewrite the integral over $\mathbf{p}$ in terms of the variable $\omega = E_p$. The integral is then of the form (4) with $F(\omega)$ zero for $\omega < m$; $F(\omega)$ will depend on x_1 and x_2. The theorem applies directly; we see that the amplitude cannot be zero for any finite interval of time. In particular, it cannot be zero outside the light cone of x_1. In other words, there is an amplitude for particles to travel faster than the speed of light and no arrangement of superposition (with only positive energies) can get around that.

Therefore, if t_2 is later than t_1 we get contributions to the amplitude from particles traveling faster than the speed of light, for which x_1 and x_2 are separated by a spacelike interval ('spacelike-separated').

Now with a spacelike separation the order of occurrence of U_1 and U_2 is frame-dependent: if we look at the event from a frame moving sufficiently quickly relative to the original frame, t_2 is earlier than t_1 (Fig. 2).

What does this process look like from the new frame? Before time t_2', we have one particle happily traveling along, but at time t'_2 something seemingly very mysterious happens: at point x_2, a finite distance from the original particle, the disturbance creates a pair of particles, one of which is apparently moving backwards in time. At time t'_1, the original particle and that moving backwards in time disappear. So the requirements of positive energies and relativity force us to allow creation and annihilation of pairs of particles, one of which travels backwards in time. The physical interpretation of a particle traveling backwards in time can most easily be appreciated if we temporarily give our particle a charge. In Fig. 2b, the particle travels from x_1 to x_2, bringing, say, positive charge from x_1 to x_2, yet since x_2 occurs first it is seen as negative charge flowing from x_2 to x_1.

In other words, *there must be antiparticles*. In fact, because of this frame-dependence of the sequence of events we can say that one man's virtual particle is another man's virtual antiparticle.

To summarize the situation, we can make the following statements:

(1) Antiparticles and pair production and destruction must exist.

(2) Antiparticle behavior is completely determined by particle behavior.

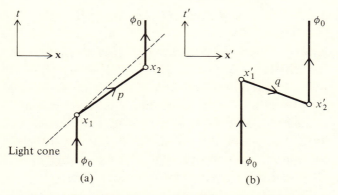

Fig. 2 The same process viewed from two different frames. (a) Original frame ($t_2 > t_1$); (b) moving frame ($t_2' > t_1'$).

We will elaborate on the second point in detail below; for now, let the following suffice. If we reversed the sign of x, y, z and t then a particle initially traveling forward in time would find itself traveling backwards in time. If we define P as the parity operator which changes the sign of the three spatial directions, T as the time reversal operation which changes the direction of the flow of time, and finally C as charge conjugation which changes particles to antiparticles and vice versa, then operating on a state with P and T is the same as operating on the state with C, that is $PT = C$.

RICHARD P. FEYNMAN

Spin-zero particles and Bose statistics

Next I would like to study the sizes of amplitudes for different processes. This will lead us along a new direction in which we will get a clue about our second subject, the connection between spin and statistics. The central idea is that if we start with any state and act on it with any set of disturbances, then the probabilities of ending up in all possible final states must add up to one.

We will first look at a nonrelativistic example and then compare it to the relativistic case. Let us suppose there is a particle initially in a state ϕ_0 and that it is acted upon by a disturbance. We want the probability of being in a given final state, calculated in perturbation theory. The amplitude that the particle is in ϕ_0 after the disturbance is given by (3); from this the probability of not doing anything is

$$\text{Prob}_{\phi_0 \to \phi_0} = 1 - 2\,\text{Re} \int d^3\mathbf{x}_1\, d^3\mathbf{x}_2$$
$$\times \int \frac{d^3\mathbf{p}}{(2\pi)^3 2E_p} b^*(\mathbf{x}_2)$$
$$\times \exp\Big\{ -i\big[E_p(t_2 - t_1) $$
$$- \mathbf{p} \cdot (\mathbf{x}_2 - \mathbf{x}_1)\big]\Big\} a(\mathbf{x}_1), \quad (5)$$

using $|1 + \alpha|^2 = 1 + \alpha + \alpha^* + \cdots = 1 + 2 \operatorname{Re} \alpha + \cdots$.

The amplitude that the particle is in state ψ_p after the disturbance is

$$\operatorname{Amp}_{\phi_0 \to p} = -i \int d^3\mathbf{x} \, \psi_p^*(\mathbf{x}) U(\mathbf{x}) \phi_0(\mathbf{x}). \quad (6)$$

Notice that in $\operatorname{Amp}(\phi_0 \to \phi_0)$ we kept terms of order U^0 and U^2 and ignored higher order terms. Here we have only a term of order U^1 and ignore terms of order U^2 and higher to get $\operatorname{Prob}(\phi_0 \to p)$ to order U^2. The probability is

$$\operatorname{Prob}_{\phi_0 \to p} = \left| -i \int d^3\mathbf{x} \, \psi_p^*(\mathbf{x}) U(\mathbf{x}) \phi_0(\mathbf{x}) \right|^2. \quad (7)$$

The total probability must be 1:

$$\operatorname{Prob}_{\phi_0 \to \phi_0} + \int \frac{d^3\mathbf{p}}{(2\pi)^3 2 E_p} \operatorname{Prob}_{\phi_0 \to p} = 1. \quad (8)$$

From this, we get a relation between the two processes: the process of scattering into another state, and the process of scattering twice ending up in the original state. Diagrammatically, this is shown in Fig. 3. It is not too much work to show

13

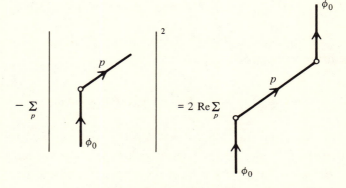

Fig. 3 A diagrammatic identity that must be true if the total probability is to be one.

that this relation is indeed satisfied for an arbitrary potential $U(\mathbf{x}, t)$.

Let us move on to the relativistic case, for spin-zero. Now we have a problem. In addition to the diagrams above, we need to allow for the fact that the intermediate state can be an antiparticle; in other words, we must add a diagram like Fig. 2b. To the total probability we must add twice the real part of this diagram. We have to find something else that cancels the contribution to the total probability of this new diagram so that the total probability remains one.

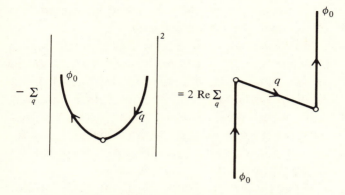

Fig. 4 A diagrammatic identity, for spin-zero particles, involving antiparticles.

A clue to the mystery is that we can make an observation, shown in Fig. 4, which is analogous to Fig. 3. This relation is not supposed to be self-evident, but if we calculate the two amplitudes we find that it is true.

The new diagram, on the left hand side of Fig. 4, forced on us by relativity, is related to the diagram where a pair is created, with the particle in the state ϕ_0. Notice that it makes a negative contribution to the total probability. So if we could introduce the diagram on the left hand side of Fig. 4 into the calculation of the total probability, the total probability would turn out to be one and we would have the problem solved.

However, simply including this diagram makes no sense, for a couple of reasons. First, the diagram on the left hand side of Fig. 4 starts from a different initial state (the vacuum rather than ϕ_0); and, second, there seems to be no reason to restrict ourselves to pair creation with the particle in state ϕ_0–any particle state is possible. We get the correct answer, but for the wrong reason.

What I have told you so far is the truth but not the whole truth. We have neglected several diagrams, and when it is all put together we will get an important feature of Bose statistics: that when a particle is in a certain state the probability of producing another particle in that state is *enhanced*.

Let us take one step back: instead of starting with a particle in ϕ_0 let us start in the vacuum V (i.e. the no-particle state), and examine our familiar idea that the total probability must be one. In the nonrelativistic case this would have been a trivial exercise: starting with no particles nothing could happen, and the probability of nothing happening would be one. In the relativistic case, on the other hand, we have seen that pair creation and annihilation must be included. Because of this, the

The reason for antiparticles

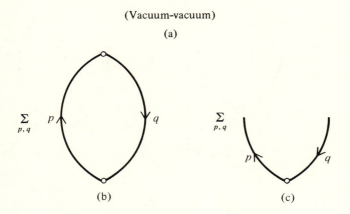

(Vacuum-vacuum)

(a)

(b)　　　　　　　　(c)

Fig. 5 Processes starting in the 'no particle state', i.e. the vacuum.

disturbance can create and annihilate pairs of particles. It is not difficult to see that, to lowest order in perturbation theory, three diagrams are important, as shown in Fig. 5. The first diagram represents nothing happening: the vacuum remains the vacuum throughout the disturbance. The second diagram is a sum of vacuum to vacuum processes, summed over all the possible intermediate particles. In the third diagram a pair is produced.

As usual, the total probability for something to happen is one. So in terms of the diagrams in

17

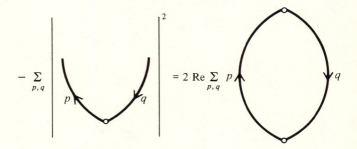

Fig. 6 A diagrammatic identity where the initial state is the vacuum.

Fig. 5, $1 = |5a + 5b + \cdots|^2 + |5c + \cdots|^2 + \cdots$, which gives the relation shown in Fig. 6.

Returning now to processes where initially we have a particle in state ϕ_0, we must include pair creation and annihilation. We get a total of six diagrams, as shown in Fig. 7. The first four restore the system to its original state, and the remaining two alter the state of the system.

We have seen in the nonrelativistic case that the probabilities from Fig. 7b and e cancel (see Fig. 3) so those from Fig. 7c, d and e must also cancel. At first sight, comparing these diagrams with those in Fig. 5, it seems as though Fig. 7d and f should cancel as did Fig. 5b and c, since they differ only by a 'spectator' particle which is irrelevant (or so it seems). We would then be left with Fig. 7c, which

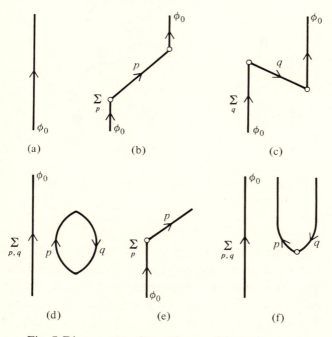

Fig. 7 Diagrams starting with a particle in the state ϕ_0.

is the problem we ran into immediately when we began the relativistic case: what cancels this contribution to the probability?

The resolution is subtle and beautiful: the 'spectator' in Fig. 7f is far from irrelevant! Consider the special case of Fig. 7f where the state p is

19

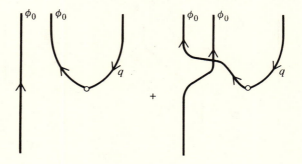

Fig. 8 One of the diagrams from Fig. 7f, with the exchange diagram.

the initial state ϕ_0. Then we have initially one particle in the state ϕ_0, and finally two particles in this state and one antiparticle in the state q. How can we be sure which of the final particles is the initial one, and which arose from the pair creation? The answer is that we cannot. In other words, we must include an extra diagram; Fig. 8 shows one of the diagrams contributing to Fig. 7f, as well as showing this so-called 'exchange' diagram.

It is this additional exchange possibility which resolves our problem. The two diagrams of Fig. 8 interfere constructively, and this extra contribution to the total probability cancels the negative contribution (see Fig. 4) from Fig. 7c.

20

The reason for antiparticles

So let me summarize the situation. We've added a few extra diagrams to account for the fact that pair production can occur; in particular we've had to add the diagram in Fig. 7c. We discovered that when we try to check the sum of the probabilities, this diagram (Fig. 7c) makes a negative contribution to the total probability, which must cancel something. What it cancels is an extra probability for producing, in the presence of a 'spectator' particle, the special particle–antiparticle pair where the newly produced particle is in the same state as the 'spectator'.

This enhanced probability is a very profound and important result. It says that the mere presence of a particle in a given state doubles the probability to produce a pair, the particles of which are in that same state. If there are n particles initially in that state, the probability is increased by a factor $n + 1$. This can obviously become very important! This is a key feature of Bose statistics, which makes the laser work, among other things.

As another example, let's look at some higher order vacuum-to-vacuum diagrams. Suppose the disturbing potential acts four times, producing and annihilating two particle-antiparticle pairs, as in Fig. 9a. Now suppose you compared that to what would happen if you produced the pairs and each

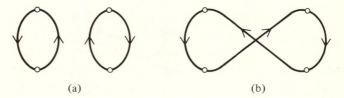

Fig. 9 (a) Two pair productions with no exchange.
(b) Two pair productions with an exchange.

particle didn't annihilate with the antiparticle it was created with, but with the other antiparticle. You would get a diagram like Fig. 9b. These two amplitudes add to make a contribution to the vacuum-to-vacuum amplitude. It is very simple and it makes Bose statistics.

Bose statistics, as a matter of fact, is not so very mysterious. The fact that the amplitude is to be added when two identical particles going A, B to A', B' arrive A to B' and B to A' instead of A to A' and B to B', seems very natural, for it appears to be merely a special case of the general quantum mechanical principle: if a process can occur in more than one alternative way, we add an amplitude for each way. Again, if particles arise from the quantization of a classical field (such as the electromagnetic field, or the vibration field of a crystal) the correspondence principle requires Bose particles if intensity correlations are to be correct, such as in the Hanbury Brown Twiss Effect.* More simply the field mode harmonic oscillators, when quantized, automatically imply a representation as Bose particles.

What we will find out later is that for fermions, particles with half-integral spin, unexpected minus signs arise. In the case of Fig. 9, for example, each loop gives the amplitude a minus sign. Therefore Fig. 9a has two minus signs, whereas Fig. 9b (which has only one loop) has one minus sign, so the amplitudes subtract and you get Fermi statistics. We are going to have to understand why with spin $\frac{1}{2}$ there is a minus sign for each loop. The key is that there are implicit rotations by 360°, as we shall see.

THE RELATION OF PARTICLE AND ANTIPARTICLE BEHAVIOR

Before we talk about fermions, I would like to return to explain in a bit more detail the relationship between particle behavior and antiparticle behavior. Of course, the antiparticle

---

* R. P. Feynman (1962). *Theory of Fundamental Processes*, pp. 4–6, W. A. Benjamin.

behavior is completely determined by the particle behavior. Let me analyze this more carefully in the simplest case of spin-zero and scalar potentials U. We have seen that for $t_2 > t_1$ the amplitude for a free particle of mass m to go from x_1 to x_2 is

$$F(2,1) = \int \frac{d^3\mathbf{p}}{(2\pi)^3 2E_p}$$

$$\times \exp\left\{ -i\left[E_p(t_2 - t_1) - \mathbf{p} \cdot (\mathbf{x}_2 - \mathbf{x}_1) \right] \right\}.$$

$$(9)$$

This formula is relativistically covariant, so for spin-zero we may take a, b constant in (5). We want to know what the amplitude is for $t_2 < t_1$.

For $t_2 < t_1$ and spacelike separation, the answer is easy: the amplitude is still $F(2,1)$. This is because we know $F(2,1)$ is correct in the spacelike region for $t_2 > t_1$, but if we look at such a process in a different frame, it must always be spacelike but we can have $t_2 < t_1$. In that frame we would get the same amplitude–it can't depend upon which frame we're in–and when we try to write $F(2,1)$ in terms of the transformed frame's coordinates we get the *same formula* because $F(2,1)$ is relativistically covariant. So $F(2,1)$ is the correct formula for the amplitude in either the forward light cone or in the spacelike region. What about the backward light cone?

The other piece of information we need is that for $t_2 < t_1$ we are still propagating only positive energies. Therefore in this region we must be able to write the amplitude in the form

$$G(2,1) = \int_0^\infty e^{+i\omega(t_2-t_1)}\chi(\mathbf{x}_1,\mathbf{x}_2,\omega)\,d\omega, \quad (10)$$

where χ is some function we want to determine. The reason for the change of sign in the exponential is as follows. We are creating waves at x_1, which we insist contain only positive energies or frequencies as we leave the source. In other words, the time dependence must be $\exp(-i\omega\,\Delta t)$ with $\omega > 0$. Here Δt is the time away from the source, which must be positive. For $t_2 > t_1$ the waves have existed for time $\Delta t = t_2 - t_1$; for $t_2 < t_1$ the waves have existed for time $\Delta t = t_1 - t_2$.

So for $t_2 < t_1$, whether in the past light cone or the spacelike region, we must be able to write the amplitude in the form of (10). This means that when $t_2 < t_1$ in the spacelike region, we could use either (9) or (10) to obtain the amplitude. This is going to determine G in that region, and it will then extrapolate uniquely for all $t_2 < t_1$.

RICHARD P. FEYNMAN

For $t_2 < t_1$ and $\mathbf{x}_1$ and $\mathbf{x}_2$ spacelike separated, we have an expression (9) which is a sum of negative frequencies. The question is, can we also express it as a function of positive frequencies alone? Ordinarily you can't do it. It's magic, but for this particular function which is relativistically invariant it is possible. Let me show you why.

First, for $t_1 = t_2$, $F(2, 1)$ is real. In that case the exponential is just $\exp[\mathrm{i}\mathbf{p} \cdot (\mathbf{x}_2 - \mathbf{x}_1)]$ and the imaginary part is an odd function integrated over an even domain, which is zero. But if F is real for $t_1 = t_2$ then it must be real for any t_1 and t_2 with spacelike separation by relativistic invariance: a moving observer would calculate the same real amplitude, yet to him $t_2 \neq t_1$. Since it is real it is equal to its complex conjugate, which has the opposite-sign time dependence. So a solution for $G(2, 1)$ is the complex conjugate of $F(2, 1)$:

$$G(2, 1) = \int \frac{\mathrm{d}^3\mathbf{p}}{(2\pi)^3 2E_p}$$
$$\times \exp\left\{ +\mathrm{i}\left[E_p(t_2 - t_1) - \mathbf{p} \cdot (\mathbf{x}_2 - \mathbf{x}_1)\right]\right\}.$$

$$(11)$$

This has the correct form: it propagates only positive energies. This must be the unique solution, for

no function of type (10) can differ from this solely
in the backward light cone, by theorem (4). So if t_2
is above t_1 in the forward light cone, the answer is
equation (9); if t_2 is below t_1 in the backward light
cone the answer is equation (11); and in the
intermediate region where t_1 and t_2 are spacelike
separated the answer is either (9) or (11)–they're
equal!

We started by knowing something in one region
of spacetime, and, just by supposing that it is
relativistically invariant, we were able to deduce
what happens all over spacetime. That's not so
mysterious. If we knew something in just one
region of a four-dimensional Euclidean space, but
knew its rotational transformation properties (in
our example, the function is invariant) we could
rotate our region in any direction and watch things
change in some well-defined way; then we could
work things out all over our Euclidean four-space.
Here we have four-dimensional Minkowski spacetime, x, y, z and t, which is a little different–but
not that much; we can still do it. The difficulty
with Minkowski space is that there is a kind of
no-man's land where t_2 is outside the light cone of
t_1: the Lorentz transformations can't really move
through there. But we have obtained the correct
continuation across this spacelike region because

supposing the energies are always positive limits the solution. In other words this operation *PT* which changes the sign of everything is really a relativistic transformation, or rather a Lorentz transformation, extended across the spacelike region by demanding that the energy is greater than zero. So it is not so mysterious that relativistic invariance produces the whole works.

Spin $\frac{1}{2}$ and Fermi statistics

So that was spin-zero, and now I would like to do spin $\frac{1}{2}$ and see what happens. If you have a spin $\frac{1}{2}$ state and you rotate it about, say, the z-axis by an angle θ, then the phase of the state changes by $e^{-i\theta/2}$. There is a whole mass of group theoretic arguments to prove this sort of thing which I won't go into now, although it's a lovely exercise. The point is that if you rotate by 360° then you end up multiplying the wavefunction by (-1). At this point all attempts to do anything by instinct fail, because this result is hard to understand. How can a complete 360° rotation change anything? One of the hardest things now will be to keep track of whether you've made a 360° rotation or not, i.e. whether you should include the minus sign or not. In fact, as we shall see, the mysterious minus signs in the behavior of Fermi particles are really due to unnoticed 360° rotations!

> Dirac had a very nice demonstration of this fact–that rotation one time around can be distinguished from doing nothing at all.* In fact, it's rotation twice around that is about the same as doing nothing. I'll show you something you can find dancing girls doing! Here–I am going to rotate this cup (see photograph sequence overleaf), re-member which way, all the way around until you can see the mark again, and now I have rotated 360°, but I'm in trouble. However, if I continue to rotate it still further, which is a nervy thing to do under the circumstances, I do not break my arm, I straighten everything out. So two rota-tions are equivalent to doing nothing, but one rotation can be different, so you have to keep track of whether you've made a rotation or not, and the rest of this talk is a nerve racking attempt to try to keep track of whether you've made a rotation or not†.

* For this demonstration due to Dirac, his famous scissors demonstra-tion, see R. Penrose and W. Rindler (1984). *Spinors and Space–time*, vol. 1, p. 43. Cambridge University Press.

† This was taken verbatim from Feynman's lecture.

I'll mention something else, just as an example, to give you an idea of the nature of the formulas that occur—it's typical to have half-angle formulas in this work. For example, suppose you have an electron and you know that the spin is $+\frac{1}{2}$ along the z-axis. Then what is the probability that if you make a measurement of spin along another axis, call it the z'-axis, that the spin will be $+\frac{1}{2}$ along this new axis? If the angle between the two axes is θ, then the answer is

$$\text{probability} = \cos^2\frac{\theta}{2} = \frac{1 + \cos\theta}{2}$$

$$\text{amplitude} = \cos(\theta/2) = \sqrt{[(1 + \cos\theta)/2]}. \quad (12)$$

The reason for antiparticles

Now we are going to study amplitudes in a spin $\frac{1}{2}$ theory with a scalar coupling. This means the disturbance, U, will be as simple as possible so that the spin parts of the amplitudes arise from the particles themselves, not from the disturbance, which will make the analysis easier. We will get formulas like the half-angle formula above, except with a relativistic modification. Here we go.

If we have a particle of mass m we know that the energy and momentum must satisfy:

$$E^2 - p^2 = m^2. \tag{13}$$

m^2 is just a constant, of course, and $p = |\mathbf{p}|$ is the magnitude of the momentum. This means that given E, p is determined and vice versa, so we don't need two different variables. Now (13) looks like the trigonometry formula $\cos^2\theta + \sin^2\theta = 1$, except for the factor m^2 and a minus sign. We can use the hyperbolic functions rather than trigonometric functions to parametrize E and p in terms of just one variable. If we write

$$E = m \cosh \omega,$$

$$p = m \sinh \omega, \tag{14}$$

then E and p automatically satisfy (13): ω is our new variable. It's called the rapidity.

RICHARD P. FEYNMAN

Suppose we have a particle at rest in a given spin state and that the disturbance puts the particle into a state with momentum p. The initial momentum four-vector is $p_1 = (m, 0, 0, 0)$, and the final one is say $p_2 = (E, p, 0, 0)$ with E and p as in (14). The amplitude for this scattering process is given by a sort of half-angle formula analogous to (12); up to irrelevant factors it is

$$A_{\text{scatt}} \propto \cosh(\omega/2). \qquad (15)$$

In analogy to the spatial rotation case above, we can write this, again up to irrelevant factors, as

$$A_{\text{scatt}} \propto \sqrt{(\cosh \omega + 1)} \propto \sqrt{(E + m)}. \qquad (16)$$

We can uniquely write this amplitude in a relativistically covariant way by noting that $p_1 \cdot p_2 = Em$, where $p_1 \cdot p_2$ is the dot product of the two four-vectors. The amplitude can therefore be written:

$$A_{\text{scatt}} \propto \sqrt{(p_1 \cdot p_2 + m^2)}. \qquad (17)$$

The power of rewriting it in a relativistically covariant way is that this amplitude, which we came up with in a special case, is now valid for any p_1, p_2. We are going to use it to derive the amplitude for pair production. Suppose we choose $p_1 = (m, 0, 0, 0)$ as before, but $p_2 = (-E, -p, 0, 0)$. This negative energy state represents an antiparticle, of course. Now $p_1 \cdot p_2 = -Em$ and we get:

$$A_{\text{pair}} \propto \sqrt{(-mE + m^2)} \propto i\sqrt{(E - m)} \quad (18)$$

as the amplitude for pair production.

Using these results, we are going to modify our discussion of total probabilities above to the case of spin $\frac{1}{2}$, and we will see that we are forced to invoke the Pauli exclusion principle. The discussion is quite similar to the spin zero case, so I will concentrate mainly on the difference between spin-zero and spin $\frac{1}{2}$.

If we study processes starting from the vacuum, the spin-zero discussion carries over directly and we get the relation shown in Fig. 6.

Let us now study processes starting from a particle in the state ϕ_0, which we now take to be a particle at rest. We get the same six diagrams as in Fig. 7, but this time the amplitudes among related diagrams obey drastically different relations.

For the total probability, we are interested in the real part of Fig. 7b, c, and d, and the absolute square of Fig. 7e and f. Let us start with Fig. 7b. In this process the particle scatters into the state p

at x_1, and propagates to x_2, where it scatters back into the state ϕ_0. From (17) we know the scatterings give a factor $\sqrt{(E + m)}$ each, so the amplitude for Fig. 7b is:

$$A_b = - \int \frac{d^3\mathbf{p}}{(2\pi)^3 2E_p} (E_p + m)$$

$$\times \exp\left\{ -i\left[E_p(t_2 - t_1) - \mathbf{p} \cdot (\mathbf{x}_2 - \mathbf{x}_1) \right] \right\},$$

$$(19)$$

where the minus sign comes from the factors $-i$ at each vertex.

The probability for the scattering process in Fig. 7e is given by the absolute square of (16), so summing over momenta it turns out that (16) and (19) imply that the relation shown in Fig. 3 holds for spin $\frac{1}{2}$ particles as well.

We must now be careful to obtain the correct expression for Fig. 7c. It must be an expression with negative frequencies equal to (19) when $t_2, \mathbf{x}_2$ and $t_1, \mathbf{x}_1$ are spacelike separated. But (19) evidently equals $-[m + i(\partial/\partial t_2) F(2, 1)]$ (see (7)), which equals $-[m + i(\partial/\partial t_2) G(2, 1)]$ in the spacelike region so Fig. 7c must be, uniquely,

$$A_c = -\int \frac{d^3 \mathbf{p}}{(2\pi)^3 2 E_p}(-E_p + m)$$

$$\times \exp\{+i[E_p(t_2 - t_1) - \mathbf{p} \cdot (\mathbf{x}_2 - \mathbf{x}_1)]\}.$$

$$(20)$$

This has been obtained by analytic continuation arguments (as in the derivation of (11)), without using (18), although the factor $(-E_p + m)$ may also be thought of as arising from two factors of A_{pair} (see (18)).

The important difference between the spin $\frac{1}{2}$ and spin-zero cases occurs at this point: the relation shown in Fig. 4 is false for fermions. To see this for the spin $\frac{1}{2}$ case we have all the necessary ingredients. From (18) we get $E_p - m$ for the necessarily positive probability of pair production; comparing this with the real part of the amplitude (20) (which has a factor $-E_p + m$ times the spin-zero amplitude) we get the relation shown in Fig. 10, which differs from Fig. 4 by a crucial minus sign.

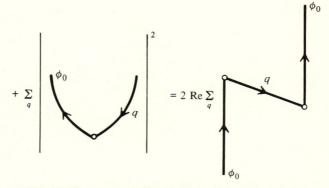

Fig. 10 The corresponding identity to Fig. 4 for spin $\frac{1}{2}$ particles.

Now, recall that in the Bose case, Fig. 7c made a negative contribution to the total probability. This meant there must be an extra, unexpected positive contribution in order for everything to balance out. The tricky business could be reduced to Fig. 7d and f with $p = 0$, and Fig. 7c. It was seen that adding the exchange diagram (Fig. 8) provided the necessary positive contribution. The net implication was that Bose statistics were needed.

In the Fermi case, in contrast, Fig. 7c makes a *positive* contribution to the total probability (as can be seen from Fig. 10) so that what we need is an extra *negative* contribution. In fact, by virtue of Figs. 6 and 10, Fig. 7c and d (with $p = 0$) exactly cancel, so we are left with the requirement that the two diagrams in Fig. 8 must exactly cancel also, in order for the total probability to be one.

From this we see that the amplitudes for diagrams which differ only by the interchange of a pair of fermions must be subtracted. It all fits together only if you say that when there is a 'spectator' particle in a certain state the probability of producing another particle in that state by new pair production is *decreased* for fermions: instead of the amplitude going up to $1 + 1 = 2$ as in the Bose case, it goes to $1 - 1 = 0$ in the Fermi case. The rule is that if you have a particle in a state you can't make another particle in that state by pair production, and the fact that the initial particle is preventing something that you expected to happen from happening, shifts the probability the other way as needed. Thus we have demonstrated for a specific example the connection between spin and statistics; that it is different for spin $\frac{1}{2}$ than it is for spin-zero. We have used relativity with quantum mechanics and have of course the formulas of the Dirac equation. We shall now continue to discuss it to obtain an even clearer idea of just why it works.

RICHARD P. FEYNMAN

ANTIPARTICLES AND TIME REVERSAL

What I would like to do now is to formulate the general rule that connects a particle to its antiparticle. We said before quite explicitly that all you need to do to work out the behavior of the antiparticle is to look at the particle 'backwards'. To be more precise, the following is true. Suppose you start with an antiparticle in some initial state of momentum $\mathbf{p}_i$, energy E_i, spin state u_i (whatever spin the particle is). Starting in that state the antiparticle could do various things. For example, if the antiparticle is charged it could emit a photon with polarization $\mathbf{a}$, momentum $\mathbf{Q}_a$, energy E_a, to end up in a final state of momentum $\mathbf{p}_f$, energy E_f, spin u_f. The amplitude for the antiparticle to do this is the *same* as the amplitude for the particle to do the reverse, namely for the particle to start off with momentum $\mathbf{p}_f$, energy E_f, spin state $(PT)u_f$, absorb a photon of polarization $-\mathbf{a}^*$, momentum $\mathbf{Q}_a$, energy E_a, to end up in a state of momentum $\mathbf{p}_i$, energy E_i, spin $(PT)^{-1}u_i$ (see Fig. 11). Hence you get the amplitude for particle behavior just by applying PT to the antiparticle behavior. Notice that PT applied to a state of momentum $\mathbf{p}$, energy E, is also a state of momentum $\mathbf{p}$, energy E. Why? Because the time reverse of the state has momen-

The reason for antiparticles

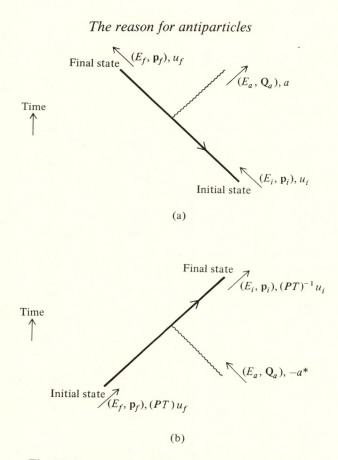

(a)

(b)

Fig. 11 A process involving (a) an antiparticle and (b) the corresponding process for the particle.

tum $-\mathbf{p}$, energy E, but then applying parity and reversing all spatial directions puts it back to momentum $\mathbf{p}$, energy E. PT does affect the polarization of the photon though, and also the spin states. Note that at one end of the process we must make the inverse transformation, i.e. a $(PT)^{-1}$ transformation. Although this sounds the same as PT, there is a subtle difference as we shall see in a minute. Hence the C that changes from particle to antiparticle is equivalent to a parity reversal P together with a time reversal T. Everything is done in the reverse order in time–for example, if you have circularly polarized light, the polarization vector is say $(e_x, e_y) = (1, i)$, the time reversed polarization is $(e_x, e_y)^* = (1, -i)$ which has the electric vector going round in the reverse direction. Then $PT(e_x, e_y) = -(e_x, e_y)^*$ and so on. $C = PT$–everything backwards in time and reversed in space. I'm not going to go through the details to prove it though.

As mentioned above, when getting the particle behavior from the antiparticle, one spin state at one end has PT applied, the other at the other end has $(PT)^{-1}$ applied. We would prefer to have the same transformation applied to both, because if the spin states u_i and u_f are the same, then so are the spin states $(PT)u_i$ and $(PT)u_f$. We will need to use this later. It turns out that there is no problem with the parity operation P, so let us choose the phases so that $P^2 = 1$, i.e. two space inversions is the same as doing nothing. What we are going to show though is that for spin $\frac{1}{2}$ particles $T^{-1} = -T$, i.e. that $TT = -1$, whereas for spin-zero $TT = +1$. That difference in sign, that extra minus sign, is where the Pauli exclusion principle and Fermi statistics come from.

THE EFFECT OF TWO SUCCESSIVE TIME REVERSALS

Why should it be that two time reversals change the sign of a spin $\frac{1}{2}$ particle? The answer is that changing T twice is equivalent to a 360° rotation. If I flipped the x-axis twice, I would be rotating through 360°, and thinking in four-dimensional spacetime; the same could be true of the t-axis too. Indeed it is true as I will demonstrate below (even without implying any relativistic relation of t and x!). Then, as we said above, rotating a spin $\frac{1}{2}$ particle by 360° multiplies it by (-1), so we find $TT = -1$. Let's show that we must have $TT = -1$ for spin $\frac{1}{2}$.

In Table 1 are listed various states, together with what you get if you apply T once, and then once

Table 1. *The effect of time reversal on various states.*

State: $|a\rangle$	Time reversed: $T|a\rangle$	Twice time reversed: $TT|a\rangle$
$|x\rangle$	$|x\rangle$	$|x\rangle$
$|p\rangle = \Sigma e^{ipx}|x\rangle$	$|-p\rangle = \Sigma e^{-ipx}|x\rangle$	$|p\rangle$
$\alpha|a\rangle + \beta|b\rangle$	$\alpha^* T|a\rangle + \beta^* T|b\rangle$	$\alpha TT|a\rangle + \beta TT|b\rangle$
Integral spin states		
$|j, m=0\rangle$	$e^{i\phi}|j, m=0\rangle$	$e^{i\phi}(e^{-i\phi}|j, m=0\rangle) = |j, m=0\rangle$
Spin $\frac{1}{2}$ states		
$|+z\rangle$	$|-z\rangle$	$-|+z\rangle$
$|-z\rangle$	$-|+z\rangle$	$-|-z\rangle$
$|+x\rangle = (|+z\rangle + |-z\rangle)/\sqrt{2}$	$(|-z\rangle - |+z\rangle)/\sqrt{2} = -|-x\rangle$	$-|+x\rangle$
$|-x\rangle = (|+z\rangle - |-z\rangle)/\sqrt{2}$	$(|-z\rangle + |+z\rangle)/\sqrt{2} = |+x\rangle$	$-|-x\rangle$

more. The first state is the state where a particle is at the point x in space; this state is written $|x\rangle$ using Dirac's notation. In between the '|' and the '$\rangle$' one puts the name of the state, or just something to label it, which in this case is the point x where the state is. Then the time reversed state is $T|x\rangle = |x\rangle$, i.e. the particle will be at the same point, no big deal. On the other hand, a particle in a state of momentum p (i.e. in a state $|p\rangle$) will time reverse into a state of momentum $-p$, but then back to $|p\rangle$ with the second time reversal.

Considering the state $|p\rangle$ shows us that T is what is called an 'antiunitary' operation. $|p\rangle$ can be made by combining states $|x\rangle$ at different positions with different phases. To get the time reversed state $|-p\rangle$ just take the states $T|x\rangle = |x\rangle$ but with the complex conjugate of the phases used to construct $|p\rangle$. So in general $T[\alpha|a\rangle + \beta|b\rangle] = \alpha^*T|a\rangle + \beta^*T|b\rangle$, i.e. for an antiunitary operation you must take the complex conjugates of the coefficients whenever you see them. Of course, if you apply T again, you take the complex conjugate of the coefficients again, and if you are very good at algebra you know that doing that is a waste of time. Now $TT|a\rangle$ must be the same physical state $|a\rangle$, but the damn quantum mechanics always allows you to have a different phase. So, by the above argument, $TT|a\rangle$ = phase $|a\rangle$ with the same phase for all states that could be superposed with the state $|a\rangle$, so that any interference between states is the same before and after applying TT. Spin-zero and spin $\frac{1}{2}$ states cannot be superposed, the two sorts of state are fundamentally different; hence the overall phase change when you apply TT can be different between the two.

What we are going to use now is that if you have a state of angular momentum $|j, m\rangle$, then $T|j, m\rangle$ = phase $|j, -m\rangle$. It must be like this for angular momentum: the time reverse of something spinning one way is the object spinning in the opposite direction. For example, with orbital angular momentum $\mathbf{L} = \mathbf{r} \wedge \mathbf{p}$, we find that since T sends $\mathbf{r} \to \mathbf{r}$ and $\mathbf{p} \to -\mathbf{p}$, then $T\mathbf{L} = -\mathbf{L}$, i.e. you get the opposite angular momentum when you apply T.

First of all consider integral spin states. There will be a state with no z-angular momentum, namely $|j, m = 0\rangle$. Applying one T this becomes the same state $|j, m = 0\rangle$ times some phase, but applying T again can only put the state back to exactly $|j, m = 0\rangle$, using the fact that T is antiunitary. So since the phase is the same for all states that can be superposed, $TT = +1$ for integral spin states.

To understand what happens with half integral spin let us take the simplest example of spin $\frac{1}{2}$. Let us try to fill out our table for just the four special states, up and down along the z-axis, $|+z\rangle$, $|-z\rangle$, and up and down along the x-axis $|+x\rangle$, $|-x\rangle$. Elementary spin theory tells us how these latter two can be expressed in terms of the $|+z\rangle$ and $|-z\rangle$ base states: one of them, $|+x\rangle$, is the in-phase equal superposition, and the other, $|-x\rangle$, is the out-of-phase equal superposition. The physically time reversed state of $|+z\rangle$ is $|-z\rangle$ and vice versa. Likewise, time reversal of $|+x\rangle$ must send us to $|-x\rangle$ within a phase.

For our first entry $T|+z\rangle$ we must have $|-z\rangle$, at least within a phase. This first phase can be chosen arbitrarily, as you can check later, so we may as well take $T|+z\rangle = |-z\rangle$. Now $T|-z\rangle$ must be a phase times $|+z\rangle$. But we cannot choose it to be simply $|+z\rangle$ because then the operation of T on $|+x\rangle$, the in-phase superposition of $|+z\rangle$ and $|-z\rangle$, will only give back the same in-phase state $|+x\rangle$ and not a factor times the out-of-phase state $|-x\rangle$, as it physically must. To make this phase reversal occur we *must*

45

take $T| - z \rangle = - | + z \rangle$, of opposite phase from what we did for $T| + z \rangle$. Now $T(T| + z \rangle) = T| - z \rangle = - | + z \rangle$ and the rest of the table can be filled out. Therefore $TT = -1$ for spin $\frac{1}{2}$, as is easily shown for any half integral spin j, where time reversal never brings us back to the same physical state. Hence combining this with the result for integral spin particles, we have $TT \equiv 360°$ rotation.

Now we come to the sign of the spin $\frac{1}{2}$ loop. You will recall that, with a potential in relativistic quantum mechanics, pairs can be produced so the probability for the vacuum (i.e. the no particle state) to remain the vacuum must be less than one. Write the amplitude for the vacuum remaining the vacuum as $1 + X$, where X is the contribution from all the closed loops drawn on the right hand side of Fig. 6. Then X must contribute a negative amount to the probability for the vacuum to remain the vacuum, which is what the identity in Fig. 6 says because the left hand side is strictly negative.

Consider the loops contributing to X. A loop is constructed by starting with an electron, for example, in a state with Dirac wavefunction u, say, and then propagating around the loop to come back into the same physical state u, and we must take the trace of the resulting matrix product,

The reason for antiparticles

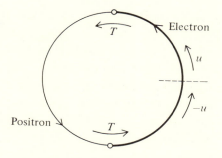

Fig. 12 A particle–antiparticle loop with the two time reversals T shown.

summing over the diagonal elements. But there is a subtlety; the same physical state could have been rotated by 360°, and indeed we see we do have that (or its equivalent, TT). Whatever frame you watch this process from, the electron at some stage changes into a positron moving backwards in time (one T), and then later turns back to an electron moving forwards in time (another T), so propagating round the whole loop you eventually come back to the state TTu; see Fig. 12.

The same TT operator will act in the boson (spin-zero) case too, but there we have $TT = +1$ so there is no problem. In the boson case, everything works out; the ordinary trace in X leads to a negative contribution. But, in the spin $\frac{1}{2}$

case, we have just found an extra minus sign. So to ensure the identity in Fig. 6 is true, to ensure that X leads to a negative contribution to the probability, we must add a new rule for half integral spin: with every ordinary loop trace we must associate an extra minus sign to compensate for the minus sign coming from $TT = -1$. If we don't put in this extra minus sign our probabilities won't add up, we won't have a consistent theory of spin $\frac{1}{2}$ particles. This sign is only consistent with Fermi statistics.

This general rule for spin $\frac{1}{2}$ loops, that for each closed loop you must multiply by -1, is why we have Fermi statistics; see Fig. 9. There is a relative minus sign between the two cases in Fig. 9a and b, because Fig. 9a has two loops, whereas Fig. 9b only has one loop. Fig. 9 thus says that when swapping two particles around you must introduce a relative minus sign, i.e. Fermi statistics!

MAGNETIC MONOPOLES, SPIN AND FERMI STATISTICS

Finally, to elucidate still more clearly the relation of the rotation properties of particles and their statistics, I would like to show you an example in which we have a spin $\frac{1}{2}$ object for which we know

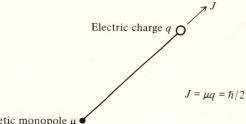

Fig. 13 A magnetic monopole μ in the presence of an electric charge q.

where the angular momentum comes from. Suppose we had a magnetic monopole in the presence of an electric charge (see Fig. 13). A magnetic monopole is something that Dirac invented, so it's appropriate to mention it in this lecture. A magnetic monopole is a source of magnetic flux in the same way an electric charge is a source of electric flux. No one has ever seen a magnetic monopole, but we can always imagine. In fact, if you just had a very long ordinary bar magnet then the magnetic flux coming out of one end would look a bit like one of these magnetic monopoles because the other end would be so far away.

Anyway, suppose we had a magnetic monopole with magnetic charge μ in the presence of an electric charge q, and we'll suppose that both these objects have spin-zero, so we don't have to worry about any intrinsic angular momentum. But these objects are in each other's presence, so you can form the Poynting vector $\mathbf{E} \wedge \mathbf{B}$ in the normal fashion. Integrating over the Poynting vector tells you what the momentum is, and, if you work it out, this composite object has an angular momentum (along the line joining the charge and pole) which is independent of how far apart the two objects are. You can work out what the angular momentum is in many ways, and I'll leave it as an exercise, but it turns out that the angular momentum is equal to μq*.

Now in quantum mechanics angular momentum must be quantized. In fact, one is only allowed to have angular momentum in multiples of $(1/2)\hbar$, so let's take the smallest value allowed, that is let $\mu q = (1/2)\hbar$, so we have constructed ourselves a spin $\frac{1}{2}$ object. Then we should find that rotating this object through 360° changes the phase by -1; let's see if it does.

---

* Perhaps the most elementary fashion for determining the angular momentum is to find the torque that must be applied to slew the axis (the line joining q and μ) around at angular velocity ω by moving the electron around a circle about the pole. The force, of course, comes from the motion of the electron in the magnetic field of the pole.

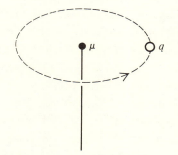

Fig. 14 Rotating the electric charge q 360° around the magnetic monopole.

Suppose the magnetic charge is fixed, and let me rotate the electric charge around it by 360° (see Fig. 14). Now there's a famous theorem that states that when you move an electric charge q through a magnetic field then the phase changes by $\exp(iq\int \mathbf{A} \cdot d\mathbf{x})$, where $\int \mathbf{A} \cdot d\mathbf{x}$ is the line integral of the vector potential $\mathbf{A}$ along the path that the electric charge follows. (That's meant to intimidate you!) In this situation, the line integral will be round the circle, but simple vector calculus tells me that I can convert the line integral of $\mathbf{A}$ into a surface integral of $\mathbf{B}$, the magnetic field, over a surface which has the circle as a boundary. Suppose I convert the line integral into an integral of

B over the upper hemisphere. The surface integral of **B** is just the flux flowing through the surface. Now the total flux emitted by the magnetic monopole is $4\pi\mu$, i.e. the integral of the flux over an entire sphere which completely enclosed the monopole would be $4\pi\mu$. Here we're only integrating over a hemisphere so we get half this, namely $2\pi\mu$. Thus the total phase change will be $\exp(2\pi i\mu q)$ and using $\mu q = \frac{1}{2}$, this works out as $\exp(i\pi) = -1$, no problem, it's absolutely right.

At this point I must digress briefly, because we are so close to an argument of Dirac's which shows that if just one monopole exists somewhere in the universe then electric charge must be quantized. The argument goes like this. Had I chosen to integrate over the lower hemisphere instead of the upper one, I would have gotten the same answer. In that case, the surface has the opposite orientation with respect to the direction of the line integral, so the phase change turns out to be $\exp(-i\pi)$, which is still (-1). But notice that if the charge q were not quantized, at a multiple of $\hbar/2\mu$, then the two different surfaces would give different answers; an inconsistency. Hence the existence of magnetic monopoles implies charge quantization, and since we believe charge is

The reason for antiparticles

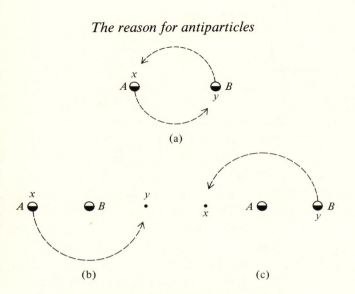

(a)

(b)   (c)

Fig. 15 Exchanging two charge–pole objects.

quantized this makes some people believe in
magnetic monopoles.

Now suppose that we have two of these things,
one charge–pole composite which I will call *A*,
and another identical one which I will call *B*.
Initially *A* is at the point *x* and *B* is at *y* (both
pointing in the same direction, say up). What
happens if I exchange them? Well, watch.

Fig. 15a shows the exchange operation. We want
to compute the phase change acquired in this

process. The only sources of phase changes are from the change in A moving round the pole in B, and the change in B moving round the pole in A. (The relative positions of the charge and pole within each composite do not change.) As seen by B, the exchange operation looks like Fig. 15b, while to A it looks like Fig. 15c. Each relative motion contributes to the total overall phase change an amount $\exp(iq\int\mathbf{A} \cdot d\mathbf{x}/\hbar)$. Since A is moving $180°$ around B, and B itself is moving $180°$ around A, there is a $360°$ rotation here. Working out the phase change by looking at the line integrals, Fig. 15b gives a line integral from x to y, but Fig. 15c gives a line integral returning from y to x–putting the two together you get the line integral around a complete closed loop around a pole just as for a $360°$ rotation and hence a factor of (-1), as we have seen before. This is exactly what you expect with Fermi statistics of course–one has a factor of (-1) when two spin $\frac{1}{2}$ objects are interchanged. (We assumed the spin-zero parts, charges and poles, obeyed Bose rules.)

SUMMARY

We've gone a long distance in great detail, but the basic ideas are the things to remember. Here's how

The reason for antiparticles

Richard Feynman (presenting
the Dirac Memorial Lecture).

it went. If we insist that particles can only have
positive energies, then you cannot avoid
propagation outside the light cone. If we look at
such propagation from a different frame, the
particle is traveling backwards in time: it is an
antiparticle. One man's virtual particle is another
man's virtual antiparticle. Then, looking at the
idea that the total probability of something happening must be one, we saw that the extra diagrams
arising because of the existence of antiparticles
and pair production implied Bose statistics for

spinless particles. When we tried the same idea on fermions, we saw that exchanging particles gave us a minus sign: they obey Fermi statistics. The general rule was that a double time reversal is the same as a 360° rotation. This gave us the connection between spin and statistics and the Pauli exclusion principle for spin $\frac{1}{2}$. That contains everything, and the rest was just elaboration.

———————

This is properly all that was in the lecture, but from talking with some of you and further thought, I should like to add some remarks that make the connection of spin and statistics still more obvious and direct. The discussion of the pole and charge objects obtained its result not through relativistic analysis of the action of two time reversals, but directly as the result of a 360° rotation. This argument can be made more general. We take the view that the Bose rule is obvious from some kind of understanding that the amplitude in quantum mechanics that correspond to alternatives must be added. What about the Fermi case?

We have noted that for half integral spin objects the sign of an amplitude might be obscure, for

360° rotations may have occurred without having been noticed.

Now the spin–statistics rule that we wish to understand can be stated for both cases simultaneously by the following single rule: *The effect on the wave function of the exchange of two particles is the same as the effect of rotating the frame of one of them by 360° relative to the other's frame.* And why should this be true? Why, simply because such an exchange implies exactly such a relative frame rotation!

We have already noticed, in the pole–charge example, that if *A* and *B* are swapped (by paths that do not exactly intersect) *A* finds *B* going around it by a 180° rotation, and *B* sees *A* going around it also by 180° in the same direction; a mutual rotation by 360°.

To verify that this is true generally, we can imagine (using an idea of David Finkelstein) the objects *A* and *B* connected at corresponding points to the ends of a ribbon or belt running between them. We can verify whether the frames have rotated relatively by looking for a 360° twist in the ribbon (when the ends *A* and *B* are exchanged the spacial location of the ribbon is approximately restored). And sure enough, swapping the objects (each moving parallel to itself, with no absolute

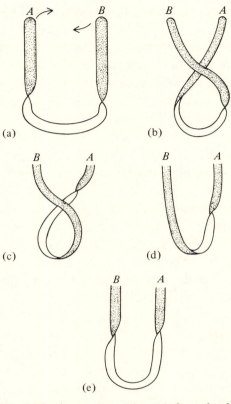

Fig. 16 In the sequence (a) to (e) the ends of the belt have been reversed in position. Note that the twist on the right-hand side in (e) comes out opposite that in (a). To restore it completely, an additional 360° turn of the right belt around the vertical would be necessary.

rotation) induces exactly such a twist in the ribbon (see Fig. 16).

Since exchange implies such a 360° rotation of one object relative to the other, there is every reason to expect the (-1) phase factor occasioned by such a rotation for exchange of half integral spin objects.

TOWARDS THE
FINAL LAWS OF PHYSICS

Steven Weinberg

I am very grateful to St John's College and to the Cambridge Mathematics Faculty for inviting me here to speak in honour of Paul Dirac. I was much in awe of him when as a student I learned of his great achievements. Later I had the privilege of meeting Dirac a few times, and I still am very much in awe of him. It's really quite a challenge to give a talk in honour of so great a man, and in planning it I felt that it would not be appropriate to speak about anything less than a great subject. I didn't think it would be fitting to tell you about the latest wrinkle in elementary particle physics that we discovered last week. So, instead, I am going to jump over all details, and speak about what is for people working in my own area of physics the greatest question of all: 'What are the final laws of physics?'

STEVEN WEINBERG

Steven Weinberg (presenting the Dirac Memorial Lecture).

Well, not quite. Much as I would like to honour Dirac by presenting a transparency on which I have written the final laws of physics, in fact I am not going to be able to do that. My real topic must necessarily be more modest. It will have to be

'What clues can we find in today's physics that tell us about the shape of the final underlying theory, that we will discover some day in the future?'

First of all, let me say what I mean by a final underlying theory. Over the last few hundred years scientists have forged chains of explanation leading downward from the scale of ordinary life to the increasingly microscopic. So many of the old questions–Why is the sky blue? Why is water wet? and so on–have been answered in terms of the properties of atoms and of light. In turn, those properties have been explained in terms of the properties of what we call the elementary particles: quarks, leptons, gauge bosons and a few others. At the same time there has been a movement toward greater simplicity. It's not that the mathematics gets easier as time passes, or that the number of supposed elementary particles necessarily decreases every year, but rather that the principles become more logically coherent; they have a greater sense of inevitability about them. My colleague at Texas, John Wheeler, has predicted that, when we eventually know the final laws of physics, it will surprise us that they weren't obvious from the beginning. Be that as it may, that's our quest: to look for a simple set of physical principles, which have about them the greatest possible sense of inevitability, and from which everything we know about physics can, in principle, be derived.

I don't know if we will ever get there; in fact I am not even sure that there is such a thing as a set of simple, final, underlying laws of physics. Nonetheless I am quite sure that it is good for us to search for them, in the same way that Spanish explorers, whey they first pushed northward from the central parts of Mexico, were searching for the seven golden cities of Cibola. They didn't find them, but they found other useful things, like Texas.

Let me also say what I *don't* mean by final, underlying, laws of physics. I don't mean that other branches of physics are in danger of being replaced by some ultimate version of elementary particle physics. I think the example of thermodynamics is helpful here. We know an awful lot about water molecules today. Suppose that at some time in the future we came to know everything there is to know about water molecules, and that we had become so good at computing that we had computers that could follow the trajectory of every molecule in a glass of water. (Neither will probably ever happen, but suppose they had.) Even though we could predict how every molecule in a glass of water would behave, nowhere in the mountain of computer printout would we find the properties of water that really interest us, properties like temperature and entropy. These properties have to be dealt with in their own terms, and for this we have the science of thermodynamics, which deals with heat without at every step reducing it to the properties of molecules or elementary particles. There is no doubt today that, ultimately, thermodynamics is what it is because of the properties of matter in the very small. (Of course, that was controversial at the beginning of the century, as you'll know if you've read a biography of Boltzmann, for example.) But we don't doubt today that thermodynamics is derived in some sense from deeper underlying principles of physics. Yet it continues, and will continue to go on forever, as a science in its own right. The same is true of other sciences that are more lively today and in a greater state of excitement than thermodynamics, sciences like condensed matter physics and the study of chaos. And of course it's even more true for sciences outside the area of physics, especially for sciences like astronomy and biology, for which also an element of history enters.

I'm also not saying that elementary particle physics is more important than other branches of physics. All I'm saying is that, because of its concern with underlying laws, elementary particle physics has a special importance of its own, even though it's not necessarily of great immediate practical value. That is a point that needs to be made from time to time, especially when elementary particle physicists come to the public for funds to continue their experiments.

I am being a little defensive about this because there is a nasty term that is applied to anyone who talks about final underlying laws. They are called reductionists. It is true, of course, that a naive reductionism can have terrible effects, nowhere more so than in the social sciences. But there is a sense in which I think we shouldn't argue the pros and cons of reductionism, because we are all reductionists today. These days, I imagine that a chemist would look with tremendous suspicion at any proposed law of chemical affinity that couldn't even in principle be derived from the properties of molecules.

This is an ancient attitude, which you can trace back to before the time of Socrates. But the realistic hope of finding a small set of simple principles that underlie all of physical reality dates back only sixty years, to the advent of the great revolution in physics which Paul Dirac put in its final form, the revolution known as quantum mechanics.

I was told in preparing this lecture that it should be pitched at the level of undergraduates who have had a first course in quantum mechanics. I suspect though that there may be a few strays in the audience who don't fit that description, so I have prepared a two-minute course in quantum mechanics for you. I have to stay within my two minutes so I am compelled to consider a very simple system. Consider a coin, and ignore all its properties such as motion and position, leaving only the question of whether it's heads or tails. Now classically the state is either heads or tails, and a classical theory would be one that would tell you when it jumps from one state to the other. In quantum mechanics, the state of the coin is not described by simply saying it is heads or tails, but by specifying a vector, the so-called 'state vector'. This state vector lives in a two-dimensional space with axes labelled by the two possible states, heads and tails (see Fig. 1). The arrow might point straight up along the tails axis, in which case you would say the coin is certainly tails, or it might point horizontally along the heads axis, in which case you would say the coin is certainly heads. In classical mechanics those are the only two possibilities. But, in quantum mechanics, the arrow (the state vector) can point in any intermediate direction. If it points in some intermediate direction then the coin is not definitely in either a heads configuration or a tails configuration. However, by looking at the coin you will force it into one of these two possibilities. That is, the result of your measurement will be one of the two possibilities, heads or tails. When you measure whether the coin is heads or tails, it will jump into one configuration or the other with a probability that depends on the angle that the arrow had initially.

The state vector can be described in terms of two components, one that I call H, the heads component, and the other called T, the tails component (see Fig. 1). H and T are called the 'probability amplitudes'. The probability of getting heads when you measure it is the square of H; and the probability of getting tails is the square of the other amplitude, T. Now, an ancient theorem of Pythagoras that you will have heard of tells us that the sum of the squares of the two amplitudes is the square of the length of the state vector. You also know that the sum of the probabilities, if you exhaust all possibilities, has to be one. This means the sum of the squares of the amplitudes has to be

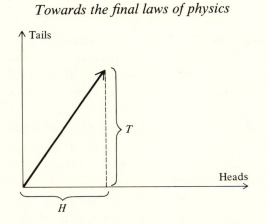

Fig. 1 A coin as an example of a simple quantum mechanical system. Probability of heads = H^2, probability of tails = T^2, so $H^2 + T^2 = 1$. The length of the state vector is $\sqrt{(H^2 + T^2)} = 1$.

one, and therefore the square of the length of this vector has to be one. In other words, the state vector must be of length one.

So in quantum mechanics a system is described by a vector of unit length, and the probabilities of an experiment giving different results are described by the squares of the components of that vector. The dynamics of the system are then described by giving a rule for how the vector rotates with time. The rule that says that in one instant the vector will rotate by a certain amount is the dynamical prescription for the system. Incidently, it's always a perfectly deterministic prescription. The state vector evolves deterministically and indeterminism only enters when you try to measure which state the coin is in.

That's all there is to quantum mechanics. Of course, for real systems it's a bit more complicated. For example, even a coin has a position, and so the state vector really lives in a much larger space: there is one direction in the space for every possible position that the coin could occupy, and when you measure its position you get a result which is a particular position with a probability that is the square of that component of the state vector. Also these are complex spaces, not real spaces, so we are really talking about complex infinite-dimensional spaces here. But this example is perfectly good for my purposes here.

Now, will quantum mechanics survive in a future final theory of physics? I would guess that it will, partly because of the enormous success that it has had over the last sixty years, but even more because of a sense of inevitability that quantum mechanics gives us. It's very interesting that, although you can read in the physics literature about efforts quantitatively to test well-established theories like general relativity, or the electroweak theory, or the theory of strong interactions, you rarely read about attempts to do quantitative tests of quantum mechanics.* The reason is that in order quantitatively to test a theory, you have to have some more general theory of which the theory you're testing is some special case. You can then ask what the more general theory predicts and then see whether the observations agree with these more general predictions, or with the special predictions of the particular theory you are interested in. Now we can make generalizations of general relativity or generalizations of the electroweak theory. These generalizations are not particularly beautiful, which is one of the reasons we believe general relativity and the electroweak theory, but nevertheless they are useful to us as straw men that we can knock down in our efforts to test general relativity or the electroweak theory.

---

* There are decisive quantitative tests of *classical* mechanics, as for instance the experiments of Aspect on spin correlations. As pointed out by Bell, classical mechanics would lead to an inequality for these correlations, which can be violated in quantum mechanics. The fact that this inequality is observed to be violated in these experiments shows that the rules of classical mechanics are badly violated in these phenomena, but it only serves to test quantum mechanics itself to an accuracy of roughly 1%.

I don't know of any generalization of quantum mechanics that makes sense. That is, I don't know of any larger logically consistent theory in which quantum mechanics appears as one special case. Usually what goes wrong when you try to generalize quantum mechanics is that, either the probabilities don't add up to one, or that you get some negative probabilities. I think it would be useful, even if you didn't believe it, to generalize quantum mechanics so that experimentalists would have something to shoot at. It may be impossible, in which case I think you would have to agree that quantum mechanics scores very high on the count of inevitability.

But quantum mechanics is not enough. Quantum mechanics is not itself a dynamical theory. It is an empty stage. You have to add the actors: you have to specify the space of configurations, an infinite-dimensional complex space, and the dynamical rules for how the state vector rotates in this space as time passes.

Increasingly, many of us have come to think that the missing element that has to be added to quantum mechanics is a principle, or several principles, of symmetry. A symmetry principle is a statement that there are various ways that you can change the way you look at nature, which actually change the direction the state vector is pointing, but which do not change the rules that govern how the state vector rotates with time. The set of all these changes in point of view is called the symmetry group of nature. It is increasingly clear that the symmetry group of nature is the deepest thing that we understand about nature today. I would like to suggest something here that I am not really certain about but which is at least a possibility: that specifying the symmetry group of nature may be all we need to say about the physical world, beyond the principles of quantum mechanics.

The paradigm for symmetries of nature is of course the group of symmetries of space and time. These are symmetries that tell you that the laws of nature don't care about how you orient your laboratory, or where you locate your laboratory, or how you set your clocks or how fast your laboratory is moving.

For example, consider rotational symmetry (or rotational invariance). This symmetry principle says that it doesn't matter how you orient your laboratory. To see how this works, let's apply it not to our coin but to something very similar, namely an electron. Just as for the coin, we'll ignore the electron's motion and just consider its spin. It's one of the peculiar facts about quantum mechanics that the space of configurations of the spin of an electron is very simple. It's a two-dimensional space, just like that of the coin (see Fig. 2). Along any axis in space, say the vertical axis, the electron spin can be either *up*, which means the electron is spinning counterclockwise around the vertical direction, or *down*, which means the electron is spinning clockwise around the vertical direction. So once again the space has two directions, but called spin-up and spin-down rather than heads and tails. If the state vector for the electron's spin is along the *up* direction, then the electron's spin is definitely up, and if the state vector is along the *down* direction then the spin is definitely down. However, it could be somewhere in between. (For example, if you happen to know that the electron is spinning clockwise around a horizontal axis, then the state is represented in our spin-up/spin-down space by a direction in-between the spin-up and spin-down axes, which correspond to spin-up and spin-down measured in the vertical direction.) Now, if you change the orientation of your laboratory by rotating it, tipping it away from the vertical direction by an angle θ (just imagine the whole room tipping over), then the state vector itself changes. In fact it rotates by an angle $\theta/2$,

74

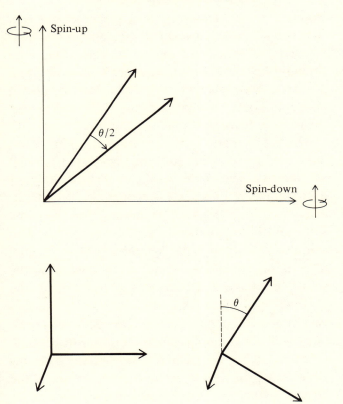

Fig. 2 The effect of rotation on an electron's spin.

which is a mathematical consequence of the fact that the electron has angular momentum (or spin) one-half, in units of Planck's constant. But even though the state vector changes, the rules that govern how the state vector varies with time do not change. That is what we mean by rotational invariance being a symmetry of nature.

There are many other symmetries which don't have to do with space and time, so-called internal symmetries. The conservation of electric charge is a consequence of such a symmetry, which physicists usually call gauge invariance. Incidentally, some of these symmetries are broken. A broken symmetry, although a symmetry of the final underlying equations, is not a symmetry of the solutions of the equations that correspond to the observable physical states. My own work has been very much in connection with broken symmetries, but I will not get into them at all in this discussion.

Now I think it's clear that any symmetry principle is a principle of simplicity. After all, if the laws of nature did depend on how you orient your laboratory, as used to be thought in the time of Aristotle, then of course they would have to contain some reference to the orientation of the laboratory relative to something else, and that would be a complication–one that we would probably find a rather ugly complication, in fact. Not having any reference to the orientation of the laboratory, the laws of nature are simpler. Nevertheless, at first sight, you might think that, even given quantum mechanics and a large number of simplifying principles of symmetry, you could still invent a huge number of extremely ugly theories that were compatible with all the symmetries and with quantum mechanics.

I think there are two reasons for being rather more optimistic than that. The first reason is that one of the symmetries we require seems to be almost inconsistent with quantum mechanics. The symmetry known as Lorentz invariance, which was part of the special theory of relativity developed by Einstein in 1905, tells us that the laws of nature do not depend on the motion of the laboratory, as long as the motion is uniform and is described in the way that Einstein described it. This symmetry is almost incompatible with quantum mechanics, so the combination of the two puts tremendous constraints on the form of any kind of dynamical theory. For example, we now know that in any such theory, for every species of particle there must be a corresponding species of antiparticle that has the same mass and spin but opposite electric charge. There is an electron, so therefore there must an anti-electron, or positron, a particle discovered in 1932. There is a proton, so therefore there must be an anti-proton, a particle discovered experimentally in 1955. Of course, this was one of Dirac's biggest hits. In his theory of 1928–30, when he tried to reconcile quantum mechanics and special relativity in his own way, Dirac found that anti-matter was inevitable. There are other similar consequences of combining relativity with quantum mechanics which are unavoidable, having to do with the behaviour of particles when you put several of them in the same state. Many of you will recall that Richard Feynman, in the first of these Dirac Memorial Lectures, demonstrated that quantum mechanics and relativity together are enough, not only to deduce the existence of anti-matter, but also to infer the behavior of particles when several of them are in the same state (the so-called spin–statistics connection).

More generally, although it is not a theorem, it is widely believed that it is impossible to reconcile quantum mechanics and relativity, except in the context of a quantum field theory. A quantum field theory is a theory in which the fundamental ingredients are fields rather than particles; the particles are little bundles of energy in the field. There is an electron field, there is a photon field, and so on, one for each truly elementary particle.

There is another reason for believing that symmetries are fundamental, and possibly all that one needs to learn about the physical world beyond quantum mechanics itself. Consider how you describe an elementary particle. How do you tell one elementary particle from another? Well you have to give its energy and its momentum, and you have to give its electric charge, its spin and a few other numbers we know about. Now if you give those numbers, that is all you can say about an elementary particle; an electron with given values of the energy, momentum etc., is identical to every other electron with these values. (In this sense elementary particles are terribly boring, which is one reason why we're so interested in them.) Now these numbers, the energy, momentum, and so on, are simply descriptions of the way that the particles behave when you perform various symmetry transformations. For instance, I have already said that when you rotate your laboratory the state vector of an electron's spin rotates by half the angle; that property is described by saying the electron is a spin one-half particle. In a similar way, though it may not be obvious to all of you, the energy of the particle just tells you how the state vector of the particle changes when you change the way you set your clock, the momentum of the particle tells you how you change the state vector when you change the position of your laboratory, and so on. From this point of view, at the deepest level, all we find are symmetries and responses to symmetries. Matter itself dissolves, and the universe itself is revealed as one large reducible representation of the symmetry group of nature.

Even so, we are still a long way from our goal of a final set of underlying laws of nature. Even assuming that we know the final symmetries, and that we believe quantum mechanics, and that we believe putting them together can only be done in the form of a quantum field theory, what we have at the very most is a framework with an infinite number of constants that still need to be determined.

Let me try and explain this in the context of a mythical universe in which the only particles are the ones that Dirac actually considered in his great work , the electron and the photon. Let's examine the following equation:

$$\mathcal{L} = -\bar{\psi}\left(\gamma^{\mu}\frac{\partial}{\partial x^{\mu}} + m\right)\psi$$

$$-\frac{1}{4}\left(\frac{\partial A_{\nu}}{\partial x^{\mu}} - \frac{\partial A_{\mu}}{\partial x^{\nu}}\right)^2$$

$$+ieA_{\mu}\bar{\psi}\gamma^{\mu}\psi$$

$$-\mu\left(\frac{\partial A_{\nu}}{\partial x^{\mu}} - \frac{\partial A_{\mu}}{\partial x_{\nu}}\right)\bar{\psi}\sigma^{\mu\nu}\psi$$

$$-G\bar{\psi}\psi\bar{\psi}\psi + \cdots . \tag{1}$$

It may not mean very much to most of you; on the other hand it means a lot to some of you! Fortunately almost all of the details are irrelevant for the points that I want to make. Let me explain briefly what all the symbols mean. $\mathcal{L}$ stands for Lagrangian density; roughly speaking you can think of it as the density of energy. Energy is the quantity that determines how the state vector rotates with time, so this is the role that the Lagrangian density plays; it tells us how the system evolves. It's written as a sum of products of fields

81

and their rates of change. ψ is the field of the electron (a function of the spacetime position x), and m is the mass of the electron. $\partial/\partial x^\mu$ means the rate of change of the field with position. γ^μ and $\sigma^{\mu\nu}$ are matrices about which I will say nothing, except that the γ^μ matrices are called Dirac matrices. A_μ is the field of the photon, called the electromagnetic field.

Looking in order at each term on the right-hand side of the equation, the first term involves the electron field twice, the next term involves the photon field twice because the bracket is squared, the third and fourth terms involve two electron fields and one photon field, the fifth term involves four electron fields, and so on. The symmetries of quantum electrodynamics give us well-defined rules for the construction of the terms in the Lagrangian, but there are an infinite number of terms allowed, with increasing numbers of fields, and also increasing numbers of derivative operators acting on them. Each term has an independent constant, called the coupling constant, that multiplies it. These are the quantities $e, \mu, G, \ldots$ in (1). The coupling constant gives the strength with which the term affects the dynamics. No coupling constants appear in the first two terms simply because I have chosen to absorb them into the definition of the two fields ψ

82

and A_μ. If there were a constant in front of the first term, for example, I would just redefine ψ to absorb it. But for all the other terms, infinity minus two of them, there is a constant in front of each term. In principle all these constants are there, and they are all unknown. How in the world can you make any money out of a theory like this?

In fact, it's not that bad. Experimentally we know that the formula consisting of just the first three terms, with all higher terms neglected, is adequate to describe electrons and photons to a fantastic level of accuracy. This theory is known as quantum electrodynamics, or QED.

In order to be able to show you in this lecture just how accurate QED is, I looked up one measured quantity, the strength of the electron's magnetic field. We can think of the electron as a little permanent magnet, whose strength as a magnet is given by a number called the magnetic moment of the electron. It is convenient to give it, not in centimetre–gramme–second units, but in what are called natural units. In these units the value one is the value that was originally obtained for the electron's magnetic moment by Dirac, in 1928. There are corrections to that value, due to the fact that the electron is surrounded with a cloud of so-called virtual photons and electron–positron pairs which the electron is continually spitting out and then reabsorbing. This has been calculated many times, first, by I believe, Schwinger, and most recently and comprehensively by Kinoshita in 1981. The result of Kinoshita's calculation, together with the current experimental value, are given below in (2).

Magnetic moment of the electron:

Kinoshita's calculation:
1.00115965246 ± 0.00000000020

Best experimental value:
1.00115965221 ± 0.00000000003.   (2)

The agreement between the two is not bad, as I think you will agree. Most of the uncertainty in the theoretical value comes from the uncertainty in the value of the electric charge of the electron, i.e. from uncertainty in the constant e in (1). So although in principle the Lagrangian for photons and electrons could be infinitely complicated, in practice only the first three terms seem to matter.

Many of us used to think we knew why the behaviour of electrons and photons is described by just the first three terms in (1). This argument, which actually goes back to work by Heisenberg in the 1930s, is one that I would have been glad to give without any reservations until about five or six years ago. The argument is based on dimensional analysis, that is on the consideration of the units or 'dimensions' of physical quantities. I will work in a system of units called physical units, in which Planck's constant and the speed of light are both set equal to one. With these choices, mass is the only remaining unit; we can express the dimensions of any quantity as a power of mass. For example, a distance or time can be expressed as so many inverse grammes. A cross-section, which would normally be a length squared, is given in terms of so many inverse grammes squared. In fact, in natural units, most observables such as cross-sections and magnetic moments have units which are negative powers of mass. Now, suppose first of all that all interactions have coupling constants which are pure numbers, like the constant e in the third term of (1). (In physical units, e has the value $\sqrt{(4\pi/137)}$ pretty nearly, and it doesn't matter what system of units you are using for mass.) Suppose that all coupling constants had no units, that they were pure numbers like e. Then it would be very easy to figure out what contribution an observable gets from its cloud of virtual photons and electron–positron pairs at very high energy E. Let's suppose an observable $\mathcal{O}$ has dimensions [mass]$^{-\alpha}$, where α is positive. (Of course, since the speed of light is one in these natural units, mass and energy are essentially the same quantity.) Now, at very high virtual-particle energy, E, much higher than any mass, or any energy of a particle in the initial or final state, there is nothing to fix a unit of energy. The contribution of high energy virtual particles to the observable $\mathcal{O}$ must then be given an integral like

$$\mathcal{O} = \int^{\infty} \frac{\mathrm{d}E}{E^{\alpha+1}} \tag{3}$$

because this is the only quantity which has the right dimensions, the right units, to give the observable $\mathcal{O}$. (The lower bound in the integral is some finite energy that marks the dividing line between what we call high and low energy.) This argument only works because there are no other quantities in the theory that have the units of mass or energy. All physicists use this sort of argument from time to time, especially when they can't think of anything else to do.

On the other hand, suppose that there are other constants around that have units of mass raised to a negative power. Then if you have an expression involving a constant C_1 with units $[\text{mass}]^{-\beta_1}$, and another constant C_2 with units $[\text{mass}]^{-\beta_2}$ and so on, then instead of the simple answer obtained above we get a sum of terms of the form

$$\mathcal{O} = C_1 C_2 \cdots \int^{\infty} \frac{E^{\beta_1 + \beta_2 + \cdots}}{E^{\alpha+1}} \, \mathrm{d}E \qquad (4)$$

again because these are the only quantities that have the right units for the observable $\mathcal{O}$. Expression (3) is perfectly well-defined, the integral converges (it doesn't add up to infinity), as long as the number α is greater than zero. However, if $\beta_1 + \beta_2 + \cdots$ is greater than α, then (4) will not be well-defined, because the numerator will have more powers of energy than the denominator and so the integral will diverge. The point is that no matter how many powers of energy you have in the denominator, i.e. no matter how large α is, (4) eventually will diverge when you get up to sufficiently high order in the coupling constants, C_1, C_2, etc., that have dimensions of negative powers of mass, because if you have enough of these constants, then eventually $\beta_1 + \cdots$ is greater than α.

Looking at the Lagrangian density in (1), we can easily work out what the units of the constant e, μ, G, etc., are. All terms in the Lagrangian density must have units $[\text{mass}]^4$, because length and time have units of inverse mass and the Lagrangian density integrated over spacetime must have no units. From the $m\bar{\psi}\psi$ term, we see that the electron field must have units $[\text{mass}]^{3/2}$, because $\frac{3}{2} + \frac{3}{2} + 1 = 4$. The derivative operator (the rate of change operator) has units of $[\text{mass}]^1$, and so the photon field also has units $[\text{mass}]^1$. Now we can work out what the units of the coupling constants are. As I said before, the electric charge turns out to be a pure number, to have no units. But then as you add more and more powers of fields, more and more derivatives, you are adding more and more quantities that have units of positive powers of mass, and since the Lagrangian density has to have fixed units of $[\text{mass}]^4$, therefore the mass dimensions of the associated coupling constants must get lower and lower, until eventually you come to constants like μ and G which have negative units of mass. (Specifically, μ has the units of $[\text{mass}]^{-1}$, while G has the units $[\text{mass}]^{-2}$.) Such terms in (1) would completely spoil the agreement between theory and experiment for the magnetic moment of the electron, so experimentally we can say that they are not there to a fantastic order of precision, and for many years it seems that this could be explained by saying that such terms must be excluded because they would give infinite results, as in (4).

Of course, that is exactly what we are looking for: a theoretical framework based on quantum mechanics, and a few symmetry principles, in which the specific dynamical principle, the Lagrangian, is only mathematically consistent if it takes one particular form. At the end of the day, we want to have the feeling that 'it could not have been any other way'.

Now, although theories like quantum electrodynamics have been immensely successful, there are several reasons for being less impressed with this success today. But perhaps first I should indicate just how successful these theories have been. I described to you the success quantum electrodynamics has had in the theory of photons and electrons, as the example of the magnetic moment of the electron demonstrates. In the 1960s these ideas were applied to the weak interactions of the nuclear particles, with a success that became increasingly apparent experimentally during the 1970s. In the 1970s, the same ideas were applied to the strong interactions of the elementary particles, with results that were so beautiful that they were generally accepted even without experimental verification, though they have been increasingly experimentally verified since then. Today we have a theory based on just such a Lagrangian as given in (1). In fact, if you put in some indices on the fields so that there are many fields of each type, then the first three terms of (1) give just the so-called standard model, i.e. the theory of strong, weak and electromagnetic interactions that we use today. It is a theory that seems to be capable of describing all the physics that is accessible using today's accelerators.

Yet we are dissatisfied. I have shown you that, in general, theories can only have a finite number of free parameters, i.e. there are at most a finite number of constants in these theories that will be pure numbers like e, because if you try to complicate the interaction you always introduce negative powers of mass into the coupling constants. One reason we are dissatisfied is that the finite number of free parameters is a fairly large number. In the present standard model I mentioned above, at the very least, assuming that there are no new particles to be discovered other than those we already know are theoretically necessary, there are seventeen free parameters that have to be chosen 'just so' in order to make the theory agree with experiment. Well, seventeen is not so many when you consider this is then going to describe all of the physics currently accessible in our laboratories. However, it is still a lot more than we would expect in an ultimate final theory–you can't say that it is obvious that these seventeen parameters should have just those values that we know experimentally they do have.

There is another reason for being dissatisfied with the standard model. This has to do with gravity. I have shown you that the condition we · require, in order not to get uncontrollable infinities when we calculate physical quantities, is that the coupling constants that describe the strength of the interaction should be dimensionless; the units should not be a negative power of mass. Now, Newton's constant, the constant which describes the strength of gravity, is not dimensionless. In physical units, we can say that Newton's constant is 10^{-10} inverse square grammes–a negative power of mass. Therefore, any theory of gravity involving Newton's constant can be expected to lead to infinities, by the same argument as given above. Many of us have tried to develop a theory of gravity that would be free of these infinities, along the lines of the kind of quantum field theory that has already been developed for the weak, electro-magnetic and strong interactions, but after many years I think most of us have had to admit failure. One promising attempt was described in the 1980 inaugural lecture of Stephen Hawking, when he entered into the Chair that he now holds, the Chair that had been held by Dirac and by Newton before him. In that lecture he grappled with the same problem of quantum gravity, and he suggested that there was one particular symmetry that could be added, called $N = 8$ supersymmetry, that might cause the infinities to cancel, and thereby lead to a theory that was finite. Today, as a result of the work of a number of theorists, it is known that the arguments which had suggested that the $N = 8$ supergravity theory is finite in low orders of perturbation theory actually break down when you go to sufficiently high orders of calculation, in the sixth order or higher I think. No one has actually shown that these infinities occur in this theory, because the calculations are too difficult, but most of us are not optimistic that this kind of theory, with or without supersymmetry, is going to lead to a finite theory of gravity interacting with everything else. Nonetheless, I should say that the spirit of Hawking's talk, of searching for the one mathematically consistent theory of all interactions, is the same one that guides me in this talk today.

Most theoretical physicists today have come around to the point of view that the standard model of which we're so proud, the quantum field theory of weak, electromagnetic and strong inter-actions, is nothing more than a low energy ap-proximation to a much deeper and quite different underlying field theory. We have two indications that nature will reveal simplicities only at an energy that is vastly higher than the energies which we can now explore. One of these indications is the fact that, if you project the coupling constants of the electroweak and strong interactions upwards in energy from the energies at which we currently measure them, you find that they all come together, they all become equal, at an energy which is about fifteen orders of magnitude greater than the mass of the proton (10^{15} GeV). Also Newton's constant of gravity, which you remember was the villain that's responsible for introducing infinities into theories of gravity, if put in physical units has the value $(10^{19} \text{ GeV})^{-2}$. This suggests that somewhere in the neighbourhood of 10^{15} to 10^{19} GeV, if we could carry out experiments at such high energies, we would find a really simple picture in which everything was unified, a picture that might even provide us with the sense of inevitability we long for.

But we can't get up there. No experiment or accelerator that is foreseeably available to the human race (certainly not in our lifetime) is going to produce energies that high. We are currently looking across a gulf of about twelve to fifteen orders of magnitude, towards a final underlying theory, with almost no help from experiment.

Now, you may wonder, since we only have a low energy approximation to a final underlying theory, a theory that will probably look nothing like the standard model that I've been talking about, why does the standard model work so well? Well, the answer is really very simple, or at least we think it is very simple. It's just that all those constants that have the units of negative powers of mass can be expected to be of the order of the same negative powers of this new fundamental energy scale. Looking back at (1), we seen that μ has units $[\text{mass}]^{-1}$, so we expect it to be of the order of $(10^{15}\text{--}10^{19} \text{ GeV})^{-1}$. Similarly, G has units $[\text{mass}]^{-2}$, so we expect it will be of the order of $(10^{15}\text{--}10^{19} \text{ GeV})^{-2}$, and so on. Those are incredibly small numbers, and therefore these constants have no appreciable effect when you compare theory and experiment for something like the magnetic moment of the electron. In other words our existing theories work well, which is certainly a reason to be happy; but we should also be sad because the fact that they work so well is now revealed as very little reassurance that any future theory will look at all like them. The standard model works so well simply because all the terms which could make it look different are naturally extremely small. A lot of work has been done by experimentalists trying to find effects of these tiny terms. There are effects like the decay of the proton, and the mass of the neutrino, which we hope will be found one day, but so far nothing has been discovered. So far, no effect except for gravity itself has been discovered coming down to us from the highest energy scale where we think the real truth hides.

That brings me to string theory. The next ten minutes are going to be excruciatingly technical; but I promise you that the technicalities which I cannot figure out how to express in non-technical language will not last until the end of the talk.

In the last two years, theoretical physicists have become intensely excited over the idea that the ultimate constituents of nature, when you look at nature on a scale of 10^{15}–10^{19} GeV, are not particles or fields but strings. I am going to discuss only one kind of string here because I want to keep the discussion as simple as I can. A string of this kind is a little loop of discontinuity in spacetime, a little glitch in spacetime that goes around in a loop. It is under tension and it vibrates, just as a string will, in an infinite number of modes. Each mode we recognize as a different species of particle. There is a lowest mode, which is the lowest mass particle, then the next mode which is the next mass particle, and so on. An interaction among elementary particles is to be seen as the joining together of these loops and then their coming apart again. Such an event can be described by giving a two-dimensional surface in spacetime, simply because as a string moves through space, it sweeps out a two-dimensional surface (a tube) in spacetime. Any particular reaction among particles is to be viewed in terms of the history of a two-dimensional surface which undergoes various splittings and rejoinings, in the course of which it absorbs those string loops that were present in the initial state and emits those string loops present in the final state. For example, a scattering process in which two particles come in and three particles go out would be described by a two-dimensional surface into which go two long tubes (representing

Towards the final laws of physics

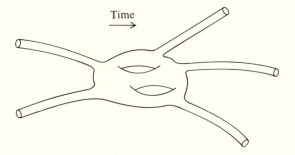

Fig. 3 A diagram corresponding to one contribution
to the scattering of two particles into three particles.

the initial particles) and out of which come three
long tubes (representing the final particles). In
between, the surface itself may have a fairly complicated topology (see Fig. 3).

The surface is described by laying down a
coordinate mesh. Since it is two-dimensional, a
point σ on the surface is identified by specifying
two coordinates which I will call σ^1 and σ^2. You
then have to be able to tell where any point on the
string is at any particular moment. To do this you
have to give a rule associating each point on the
surface, specified by $\sigma = (\sigma^1, \sigma^2)$, with a point x^μ
in spacetime. Mathematically we write $x^\mu =
x^\mu(\sigma^1, \sigma^2)$. There is also an internal metric on the

surface that describes the geometry of the surface. Just as in four-dimensional general relativity we write the metric as a position-dependent matrix $g_{\alpha\beta}(\sigma)$, although since we are only dealing with a two-dimensional surface the indices α and β here only run over the values one and two. The metric tells you how to work out the length between two points in the surface: for two adjacent points in the surface σ and $\sigma + d\sigma$, the length between them is given by $ds = \sqrt{(g_{\alpha\beta}(\sigma) \, d\sigma^\alpha \, d\sigma^\beta)}$.

The rules of quantum mechanics as formulated by Feynman tell us that if you want to know the probability amplitude (the quantity you square to get the probability for a given collision process), then you have to take a weighted average over all possible ways in which that collision process can occur. In string theory, this means that you have to sum over the histories of all two-dimensional surfaces that could produce the particular collision process. Now, each history is characterized by the two functions $x^\mu = x^\mu(\sigma)$ and $g_{\alpha\beta}(\sigma)$ described above. What you actually have to do to calculate the probability amplitude is to evaluate a numerical quantity $I[x, g]$ for each history and then add up $e^{-I[x, g]}$ for all possible surfaces. The quantity $I[x, g]$ is called the action, and is a functional of

$x^\mu = x^\mu(\sigma)$ and $g_{\alpha\beta}(\sigma)$, given by*:

$$I[x, g] = \frac{1}{2} \int \sqrt{(g(\sigma))}\, g^{\alpha\beta}(\sigma)$$

$$\times \frac{\partial x^\mu(\sigma)}{\partial \sigma^\alpha} \frac{\partial x^\nu(\sigma)}{\partial \sigma^\beta}\, \mathrm{d}^2\sigma. \qquad (5)$$

One reason that this has become so exciting is that it has been realized that these theories, for the first time, provide us with a theory of gravity which is free of the infinities that have occurred in all earlier attempts to describe gravitation. In fact, in a sense gravity was discovered in these theories. (I do know that gravity was discovered long before string theories.) These theories were originally developed in the late 1960s and early 1970s as a means of trying to understand the strong interactions of nuclear particles. It was soon found that surfaces involving long thin tubes (see Fig. 4) gave the result that in between the initial and the final particles, a pulse of radiation was being transmitted that corresponded to a massless spin-two

---

* There is actually one other term in the action, which simply serves to set the relative scale of different orders of perturbation theory.

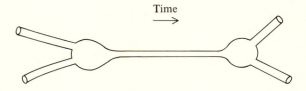

Fig. 4 A string intersection involving the emission and reabsorption of a massless spin-two particle.

particle. (Massless particles are simply particles that travel at the speed of light, and spin-two is given the same units that the electron has spin one-half.) Now, at that time that particle was a terrible embarrassment. It was known that the quantum of gravitational radiation, the graviton, has just those properties, but in the late 1960s and early 1970s string theories were supposed to deal with strong nuclear forces, not with gravity. In fact this embarrassment contributed to the eclipse of string theories after the early 1970s.

Actually, in 1974, Scherk and Schwarz suggested that string theories should be regarded as theories of gravity, but nobody took them seriously. It is only in the last few years, as a result of the work of Green, Gross, Polyakov, Schwarz, Witten, and others (including many of their younger colla-

100

borators), that physicists have begun to suspect that string theory is a very good candidate for the final laws of physics at the ultimate scale of 10^{15}–10^{19} GeV.

Now, you might feel that this prospect is not terribly attractive. After all, in order to describe the theory I had to give you an action. This action is the integral of a Lagrangian: the expression under the integral sign in (5) is the Lagrangian or energy density, just like the one I wrote down for quantum electrodynamics in (1). You might very well say, 'Who ordered that? Who said that that is the correct theory of the world? Maybe you know experimentally that that is the way ordinary strings behave, but so what? What has that got to do with ultimate elementary glitches in spacetime? Why not an infinite number of other terms? Why strings at all?'

This theory is actually rationally explicable in terms of its symmetries in a way that is not immediately obvious. The action (5) has a number of symmetries. Because of the introduction of the metric, just as in general relativity, there is a symmetry under changes in the definition of the coordinates, $\sigma \rightarrow \sigma'(\sigma)$. There is also a symmetry that is a little less obvious and is only true in two dimensions. This is symmetry under a change in the local length scale, a so-called Weyl transformation, in which we multiply the metric by an arbitrary function of position, i.e. $g_{\alpha\beta}(\sigma) \rightarrow f(\sigma)g_{\alpha\beta}(\sigma)$. And finally there is also the rather obvious symmetry under Lorentz transformations, $x^{\mu} \rightarrow \Lambda^{\mu}_{\nu}x^{\nu} + a^{\mu}$. The first two of these symmetries seem to be indispensable to the theory. Without them, you find that when you try to calculate the sums over all surfaces, the results do not make sense. Without these two symmetries, then it seems that either you get negative probabilities or the probabilities don't add up to one. In fact, there are some very subtle quantum effects that can break the symmetries, and unless you put in just the right mix of coordinates and also spin one-half coordinates (which I have not mentioned here), then there will be quantum anomalies that will spoil the symmetries. But, if you're very careful, you can get the quantum anomalies to cancel so that these symmetries remain and then the sum over the histories of these surfaces do make sense.

There is a beautiful branch of mathematics, one of the most beautiful ever invented, that deals precisely with those properties of two-dimensional surfaces that are left unchanged by coordinate and

Weyl transformations. This branch of mathematics goes right back to the work of Riemann in the first part of the nineteenth century. Many of its classic results turn out to be just what we need to understand the physics of strings. For example, to describe the topology of an arbitrary two-dimensional surface, (strictly speaking, an arbitrary oriented closed surface) all you need to do is to give the number of 'handles'. Furthermore, for a given number of handles, all you need to do to describe the geometry up to a coordinate or a Weyl transformation is to specify a finite number of parameters. These parameters then have to be integrated over when you sum over all the histories of the two-dimensional surfaces. The number of parameters here is zero for no handles, two for one handle, and is $6h - 6$ for $h \geq 2$ handles.

It is these ancient theorems that allow you to calculate the sums over the histories of the surfaces. If the symmetries weren't present you couldn't do the calculations, and they would probably give nonsense if you could. So the symmetries seem to be indispensible. Furthermore, and now this is the point, this is the punch line, the symmetries determine the action. This action, this form of the dynamics, is the only one consistent with these symmetries. (There are qualifications*, but what I am saying is close enough to the truth.) In other words, there is nothing else you could have added. There are no other possible terms consistent with these symmetries. This, I think, is the first time this has happened in a dynamical theory: that the symmetries of the theory have completely determined the structure of the dynamics, i.e. have completely determined the quantity that produces the rate of change of the state vector with time. I think it is for this reason, more than any other, that some physicists are so excited. This theory has

---

* There are about a half-dozen string theories that are consistent with the symmetries listed here, differing in the numbers of spacetime coordinates x^μ and spin one-half variables appearing in the action. Unfortunately, in all of these the number of spacetime coordinates is greater than four. One approach to this problem is to suppose that the extra coordinates are 'compactified', rolled up in a compact space of very small dimensions. However, this does not exhaust all possibilities. A more comprehensive way of searching for possible theories is to assume Lorentz invariance only for the four ordinary spacetime coordinates, and allow an arbitrary number of additional coordinates and spin one-half variables. The numbers of these ingredients and the form of the action are then to be determined by the requirement that the other symmetries (under changes in coordinates and Weyl transformations) are to be preserved despite the effects of quantum fluctuations. This program has barely been started; in particular we do not know whether there are any satisfactory theories, and if so, whether they involve any free parameters.

the smell of inevitability about it. It is a theory that cannot be altered without messing it up; that is, if you try to do anything to this theory, add additional terms, change any of the ways that you define things, then you find you lose these symmetries, and if you lose these symmetries the theory makes no sense. The sums over histories, the sums over surfaces break down and give nonsensical results. For this reason, quite apart from the fact string theory incorporates graviation, we think that we have more reason for optimism now about approaching the final laws of nature than we have had for some time.

The inventors of this theory, and those like myself who are not one of the inventors of the theory but are desperately trying to catch up with it, are now working to try and learn how to solve these theories so that we can find out what these theories say at ordinary energies and see whether they agree with the standard model. This is one point about this work that I had better explain. I said that this theory has ho free parameters in it, and it doesn't; there are no free numbers. But actually that's slightly misleading because x^μ is the position vector of a point on the string in natural string units. If instead you want to use units like centimeters, then you have to introduce a constant in equation (5). That constant is the so-called string tension; it's the conversion factor between string units and ordinary metric units. The string tension is the square of an energy which we estimate (from our knowledge of Newton's constant of gravity) to be about 10^{18} GeV, and so this is the fundamental scale for string theory. When I say 'solve the string theory', what I mean is to find out what these theories predict at much lower energies than 10^{18} GeV, where we do our experiments. The aim today is to try to find out whether the theory does in fact predict the standard model of weak, electromagnetic and strong interactions. If it does then the second question is, what does it predict for those seventeen or more parameters of the standard model? Does it tell us directly the mass of the electron, the mass of the quarks and so on? If it does, then that's it. Many of us are betting the most valuable thing we have, our time, that this theory is so beautiful that it will survive in the final underlying laws of physics.

Now, I would like briefly to address the question of what we mean by the beauty of a physical theory. In closing my lecture, I would like to enter into a short posthumous dialogue with the spirit of Dirac, who in any case I suspect is probably not very far from Cambridge. I once heard Dirac say in a lecture, to an audience which largely consisted of students, that students of physics shouldn't worry too much about what the equations of physics mean, but only about the beauty of the equations. The faculty members present groaned at the prospect of all our students setting out to imitate Dirac. Actually I partly agree with Dirac, but only partly. Beauty is our guide in theoretical physics, but it's not the beauty of the equations printed on a piece of paper that we're searching for, it is the beauty of the principles, how they hang together. We want principles that have a sense of inevitability. It doesn't matter whether the equations are more or less beautiful. For example, Einstein's general theory of relativity is characterized by a set of second-order differential equations; so is Newton's theory of gravity. From that point of view they are equally beautiful; in fact Newton's theory has fewer equations, so I guess in that sense it is more beautiful. But Einstein's theory of general relativity has a greater quality of inevitability. In Einstein's theory there was no way you could have avoided an inverse square law, under the conditions in which an inverse square law applies, namely at large distances and at slow speeds*. You can't fool around with Einstein's theory and get anything but an inverse square law at large distances without the whole thing falling apart. With Newton's theory, it would have been very easy to get any kind of inverse power you liked, so Einstein's theory is more beautiful because it has a greater sense of rigidity, of inevitability.

Nevertheless, there is another sense in which I do agree with Dirac. We flounder about in theoretical physics, very often not knowing precisely what the principles are that we are trying to apply. And without knowing what the principles are, very often the beauty of the mathematics is our best guide. In fact, very often beautiful mathematics survives in physics even when, with the passage of time, the principles under which it was developed turn out not to be correct ones. For example, Dirac's great work on the theory of the electron was an attempt

---

* This isn't strictly true, because there is one term, the so-called cosmological constant, that could be added to Einstein's theory. This term in effect produces a force that *increases* with distance. If we assume that this term is small enough not to interfere with our understanding of large bodies, like the galaxy or the whole universe, then it would have a negligible effect on the scale of the solar system. However, it must be admitted that no one today knows *why* the cosmological constant term should be this small.

to unify quantum mechanics and special relativity by giving a relativistic generalization of the Schrödinger wave equation. I think today that that point of view is generally abandoned*. These days most people think that you cannot unify special relativity and quantum mechanics except in the context of quantum field theory. (A string theory is a sort of quantum field theory.) Therefore the principles that Dirac was following have been abandoned, but his beautiful equation has become part of the stock and trade of every physicist; it

---

* For one thing, Dirac's approach works well only for spin one-half particles, like the electron. In fact, Dirac felt that it was a success of his theory that it explained *why* the electron must have spin one-half. Of course, Dirac knew that there were other particles, like the α-particle, which do not have spin one-half, and the only significant distinctive characteristics of the electron is that it is supposedly more elementary, so Dirac must have meant that all *elementary* particles must have spin one-half. This was challenged in a 1934 paper of Pauli and Weisskopf, who showed that it was perfectly possible to make a relativistic quantum theory of spin-zero particles, but it is a quantum field theory, like the quantum theory of electromagnetic fields, not a relativistic version of Schrödinger wave mechanics. (Indeed, if charged these spin-zero particles have distinct antiparticles, and since they are bosons the antiparticles clearly can't be regarded as holes in a sea of negative energy particles.) Today we know of particles like the W and Z boson and possibly the Higgs boson, that do not have spin one-half, and seem to be every bit as elementary as the electron.

survives and will survive forever. I don't know, of course, whether Dirac would think that the mathematics of string theory is sufficiently beautiful to make it likely that it will survive as part of the final laws of physics. He might agree with that, and he might not agree with that, but I don't think he would disapprove of what we are trying to do.